Hermann Weber, Frederick Parkes Weber

The spas and mineral waters of Europe

With notes on balneo-therapeutic management in various diseases and morbid conditions

Hermann Weber, Frederick Parkes Weber

The spas and mineral waters of Europe
With notes on balneo-therapeutic management in various diseases and morbid conditions

ISBN/EAN: 9783337562168

Printed in Europe, USA, Canada, Australia, Japan

Cover: Foto ©berggeist007 / pixelio.de

More available books at **www.hansebooks.com**

THE

SPAS AND MINERAL WATERS OF EUROPE

WITH NOTES ON BALNEO-THERAPEUTIC MANAGEMENT IN VARIOUS DISEASES AND MORBID CONDITIONS

BY

HERMANN WEBER, M.D., F.R.C.P.

Consulting Physician to the German Hospital and to the Royal National Hospital for Consumption, Ventnor; one of the Hon. Vice-Presidents of the British Balneological and Climatological Society; Corresponding Member of the Balneological Society of Berlin; &c.

AND

F. PARKES WEBER, M.D., M.R.C.P.

Physician to the German Hospital

LONDON

SMITH, ELDER, & CO., 15 WATERLOO PLACE

1896

PREFACE

This book is intended to supply some elementary knowledge respecting the Spas of Europe, the methods of treatment adopted there, and the diseases and morbid conditions which are most likely to be cured or ameliorated by them. Although there are already a good many works on the subject, especially in the German and French languages, as a glance at the Bibliography (see end of book) will show, a short work like the present one, in which it is hoped the most general information can easily be found, may be of use to some who have not more elaborate volumes at hand.

As much of the effect of mineral water treatment (balneo-therapeutics) cannot be separated from the effect of the external or internal use of plain water (hydro-therapeutics), the first chapter is devoted to a short consideration of hydro-therapeutics in general. Chapters II. and III. deal with the classification and action on the body of mineral waters. In the fourth chapter climate, diet, and altered mode of life are considered, as regards their share in the results obtained by spa treatment; massage and muscular exercises at spas are likewise noticed. The subjects of medical supervision, the 'after-cure,' etc., are dealt with in the fifth chapter. The next eleven chapters are given up to the uses of the different classes of waters and to notices of the different spas. Marine spas [it must be mentioned that no line has been drawn in this chapter between true marine spas and seaside localities used rather as climatic stations than for sea-bathing] are considered in the seventeenth chapter, and in the remaining chapter an attempt is made to indicate what balneo-therapeutic management is most likely to prove useful in various diseases and morbid conditions.

In describing the individual spas and mineral waters there has been necessarily a good deal of repetition, which, however, it is hoped may be found of use by those who chance to consult the book for information concerning some particular spa only. For purposes of reference also, a number of spas (some merely of local importance) have been mentioned, though they are hardly likely to be known or visited by English medical men or patients.

In the arrangement of spas of any one group, the method ordinarily followed has been to give at first somewhat detailed notes of a few of the better known members of the group, and then, in proceeding with the other members, to follow the political geographical order of Great Britain, Belgium, Germany, Austria, Switzerland, France, Italy, etc. Thus in the chapter on the simple thermal waters, Wildbad and Ragatz-Pfaefers have been taken as types of the group, and somewhat detailed accounts have been given of these two spas. The other members of the group have then been described at less detail in their political geographical order. Similarly, in the chapter on chalybeate waters descriptions of Spa, Schwalbach, and St. Moritz have been given at first; in the chapter on simple alkaline waters Vichy and Vals have been described first; in the muriated alkaline group Ems and Royat have the first places; and the earthy waters are headed by Wildungen and Contrexéville.

The most useful works of reference on the subject have been mentioned in the Bibliography at the end of the book. This, however, has no pretensions to be anything like a complete list of the immense amount of literature which has been published even in quite modern times on balneological subjects. A great amount of help has been obtained from the Royal Medical and Chirurgical Society's report on 'The Climates and Baths of Great Britain,' London, 1895; and to this work we would refer readers desiring more detailed information on the spas of Great Britain. Amongst other works found most useful we may mention those of Durand-Fardel, Seegen, Braun, Flechsig, Reimer, Leichtenstern, De la Harpe, Vintras, Macpherson, Gsell-Fels, and Valentiner.

Owing to the variety of our sources of information, and the occasional contradictory results of different analyses,[1] we are fully aware that in quoting the approxi-

[1] The variations which occur naturally in the constitution of mineral springs (from which surface water is kept out) appear to be usually only very slight. Analysis of the deposit found in pipes which supplied the ancient Roman *Thermæ* at Bath, shows that the relative proportion in the solid constituents of the Bath water must have remained unchanged from Roman times till now (see Kerr, *Bath Waters*, 7th ed., p. 64). R. Fresenius (*Veröffentl. d. Allg. deutsch. Bäder Verbandes*, 1894, p. 116)

mate amount of ingredients in mineral waters, some inaccuracies[1] must have crept in; we believe, however, that the figures will on the whole be found useful. It has been considered that the number of grammes in a litre (1,000 cubic centimetres of the water), although the first is a weight and the second a volume, may be sufficiently accurately expressed as the quantity per mille. It would be convenient if analyses of mineral waters were always expressed in this form (or else as grammes per 1,000 grammes by weight of water), which has the advantage of being international, and which, owing to the introduction of one per cent. solutions into the British Pharmacopœia, is an expression of measurement already familiar to English medical men, and with the use of decimal weights and measures in prescribing, will soon become still more so. The amount per mille need of course only be multiplied by 70 to transform it into the number of grains per imperial gallon. All temperatures are given in degrees of the Fahrenheit scale, as is usual in England.

Although climatic matters are necessarily often touched upon, the book is essentially of balneo-therapeutic, not of climato-therapeutic character. The very short statement about climatic health resorts for the important question of management after the spa treatment ('after-cure') will, we hope, be found useful.

shows that slight quantitative variations in the constitution of some German mineral waters do probably take place, and that these variations are less in thermal springs, such as the Wiesbaden Kochbrunnen, than in cold springs, such as Niederselters.

[1] The authors will, of course, be glad to see copies of recent analyses, so that errors may be corrected, should there be another edition.

It will be conceded, we trust, that the usefulness of balneo-therapeutics is not overstated, but restricted to fair limits. By consulting Chapter XVIII. (on selecting spas and mineral waters for various affections) and comparing it with the description of spas in the preceding chapters, it will be seen that complaints, the same as far as name goes, may be treated with advantage at different localities. It is hoped that the present volume may be of service to the medical man in his selection of a spa and of a locality for after-treatment to suit the requirements and convenience of individual patients, at least as far as this can be done by books.

It is, however, very important for the doctor at home to make himself, by personal visits, more intimately acquainted with the different spas and health resorts and with their local doctors. A thorough knowledge is necessary of the special features of the localities, their situation, their natural beauty and configuration, their climate and vegetation, as well as the accommodation, prevailing habits, and the society probably to be met with. The knowledge of the character and qualities of the local medical men is equally, even more important. It will be acknowledged that the amount of influence for good which the doctor at home exercises on his patients affected with chronic troubles, depends to a great degree not only on his intimate knowledge of the disease and the constitution, but on a certain sympathy, in the wider sense of the word, which arises from an insight into the mental condition and character of the patient, and which enables the doctor to put himself into accord with him, and express his advice in such a way that it induces the patient to follow it. It is therefore necessary to find, if possible, a doctor at the spa who is conscientious, intelligent, sympathetic, and firm. Shakspeare would have been the greatest physician, and amongst recent doctors Sir William Gull owed his great success to such qualities.

H. W.
F. P. W.

May, 1896.

CONTENTS

Comparative Scale of Temperatures.

Centigrade	Fahrenheit	Réaumur	Centigrade	Fahrenheit	Réaumur
80°	176°	64°	50°	122°	44°
79	174·2	63	49	120·2	39·2
78	172·4	62·4	48	118·4	38·4
77	170·6	61·6	47	116·6	37·6
76	168·8	60·8	46	114·8	36·8
75	167	60	45	113	36
74	165	59·2	44	111·2	35·2
73	163	58·4	43	109·4	34·4
72	161	57·6	42	107·6	33·6
71	159·8	56·8	41	105·8	32·8
70	158	56	40	104	32
69	156·2	55	39	102·2	31·2
68	154·4	54·4	38	100·4	30·4
67	152·6	53·6	37	98·6	29·6
66	150·8	52·8	36	96·8	28·8
65	149	52	35	95	28
64	147·2	51·2	34	93·2	27·2
63	145·4	50·4	33	91·4	26·4
62	143·6	49·6	32	89·6	25·6
61	141·8	48·8	31	87·8	24·8
60	140	48	30	86	24
59	138·2	47·2	29	84·2	23·2
58	136·4	46·4	28	82·4	22·4
57	134·6	45·6	27	80·6	21·6
56	132·8	44·8	26	78·8	20·8
55	131	40	25	77	20·0
54	129·2	43·2	24	75·2	19·2
53	127·4	42·4	23	73·4	18·4
52	125·6	41·6	22	71·6	17·6
51	123·8	40·8	21	69·8	16·8
			20	68	16

THE SPAS AND MINERAL WATERS

OF

EUROPE

CHAPTER I

HYDROTHERAPEUTICS, OR THE THERAPEUTIC USE OF PLAIN WATER

As hydrotherapeutic treatment plays a very important part in the results attained at many spas, and as the effect of many mineral water baths is nearly the same as the effect of the external application of plain water at a given temperature, it has been thought advisable to give a short account of the principles of hydrotherapeutics.

Definition, &c.

Hydrotherapeutics (Hydrotherapy) deals with the therapeutic use of simple water when taken internally or when applied externally in the form of baths, douches, &c. Hydrotherapeutic treatment may be employed in both chronic and acute affections, but for the purpose of this book it is only necessary to consider its use in chronic diseases.

History

Though known to the ancient Greeks and Romans and practised to a variable extent in modern Europe from the sixteenth century, it was Vincent Priessnitz, of Graefenberg in Silesia, who first made this kind of treatment widely known. Its too indiscriminate and energetic use, however, often led to bad results, and its more serious study by regular physicians became urgently needed. In the latter half of the present century much labour has been devoted to the scientific study of the subject, with the result that hydrotherapeutic treatment now rests on a firmer scientific basis, as the works of Winternitz, Hayem, and many other writers abundantly show.

Internal use of water

The internal use of plain water.—The amount of liquid taken habitually varies much in different individuals, partly owing to individual tendencies and mere habit, partly owing to mode of life. Abundant water drinking probably in most cases leads to increase in the fluids secreted by the body, and for a time at least to increased excretion of urea and the waste products of tissue metabolism, the tissues and the blood itself being, so to speak, washed out by the treatment. In cases of old standing valvular heart disease, especially in imperfectly compensated mitral disease, and in obese people with weakly acting hearts it is often important to diminish the fluid part of the food; in other cases excessive water-drinking may lead to dyspeptic troubles. Nevertheless, increase in the amount of water taken internally may be of service as part of the treatment for gout, tendency to urinary gravel or gall-stones, and in constipation from insufficient intestinal secretion.

A considerable part of the results obtained from courses of mineral waters is really due simply to the increase in the amount of water drunk. Small draughts of cold water act as a stimulant to the musculature of the stomach, and probably by stimulation of the pneumogastric nerve, the frequency of the heart's action is temporarily diminished. Warm water as taken internally at indifferent thermal springs is more rapidly absorbed, and subtracts less heat from the body than cold water. It does not give rise to the unpleasant shock sometimes caused by the latter.

Methods of external application

The external use of plain water.—The modes of application are very various, comprising the ordinary full bath at different temperatures, hip-baths, wave-baths, and baths in running water, wrapping in wet sheets, shower baths, affusions and all kinds of douches. The temperature of the douches can be varied during the application (the so-called 'Scotch douche' or 'alternating douche'). Though the modes of application are manifold, in practice it has been found that when good results are possible, they can usually be obtained by the judicious use of a very limited number of appliances. 'Electric baths' combine hydrotherapeutic and electrical treatment. In the ingenious 'hydro-electric douche,' which was lately investigated by Paul Guyenot in France, the incident stream of water is made to serve as the anode or the kathode of the electric current, which is applied simultaneously with the douche.

Reaction of the body to cold and heat

The results of hydrotherapeutic treatment by all these means are in great measure due to the natural reaction of the body to cold and heat; water is generally preferred to air for this purpose, because its greater specific heat and greater co-efficient of heat-conductivity renders it more active in bringing about the reaction. Human beings are particularly susceptible to hydrotherapeutic effects, because the skin, unlike that of most warm-blooded animals, is unprotected by any natural covering. The clothes, moreover, by which the body is usually protected, render the skin still more sensitive to changes of temperature, since they form a kind of habitual thermal zone about the body, the temperature of which has been shown by Winternitz to remain fairly constant at about 89°·6 F. (32° C.) A bath must therefore have a temperature of some degrees below or above 89°·6 F. in order that a decided reaction can be obtained.

Hydrotherapeutic reaction for cold

When a man jumps into a cold bath, the cold produces at first a disagreeable impression. He shivers, and after an almost involuntary pause in breathing, inspires very deeply. Owing to the contraction of the cutaneous blood vessels, the skin is pale and the contraction of the unstriped muscle fibres gives rise to the phenomenon of goose skin. These effects may give place to the 'reaction' whilst he still remains in the bath, or only when he comes out of it. The pallor of the skin changes to slight reddening, accompanied by an agreeable subjective sensation of warmth, easy breathing, and a feeling of comfort and capability for exertion.

The rapidity and degree of this 'hydrotherapeutic reaction for cold' varies very much in different individuals. It is delayed in the weak and feeble, but takes place rapidly in the robust and strong, especially if these have been in the habit of taking cold baths. The reaction in a given case depends on the temperature of the water, the length of the application, and in the case of a douche, on the force with which it is applied; it is greatly assisted by voluntary movement and friction. The best reaction with the least loss of heat is obtained by cold applications of short duration.

Mechanism of the reaction

The vascular and respiratory phenomena following on the application of cold water to the surface of the body have been experimentally shown to be mainly, if not entirely, due to nervous reflex action; moreover, in the case of men with partial paralysis and anæsthesia, the vascular phenomena have been found diminished or absent in the paralysed limbs. That the respiratory phenomena are not altogether voluntary is shown by similar reflex movements being obtained through stimulation of the skin in animals rendered insensible by chloral (Roehrig).

The initial pallor of the skin is due to contraction of the cutaneous blood-vessels, and must be interpreted as a reflex attempt on the part of the organism to hinder excessive loss of heat, or, at least, to moderate it until increased heat production in the body can be established to counterbalance the increased loss. When the reaction sets in, the cutaneous blood-vessels become dilated, giving rise to redness of the skin and a subjective sensation of warmth. The blood is the great carrier and distributor of heat through the body, and it is probable that, corresponding with the cutaneous pallor, the internal blood-vessels become dilated and the central temperature rises slightly; whereas, when the surface vessels during the reaction dilate, the skin feels warmer, and, the internal parts of the body being less supplied with blood, the central temperature undergoes a slight fall.

Results of cold water applications

Increased heat production in the body requires increased combustion in the tissues, and this is evidenced, as it is during muscular exercise, by increase in the amount of carbonic acid gas expired. It is probable also that judicious hydrotherapeutic treatment renders the combustion in the body more complete, for increase of urea and diminution of uric acid has sometimes been observed in the urine. The increased flow of urine following the bath is certainly not due to the very small quantity of water estimated as absorbed during a bath through the skin, but is doubtless due to increased blood pressure and circulation in the kidneys.

By cold water treatment the heart's action should be strengthened, the appetite increased, the digestion of food aided, and the movements of the bowels rendered more active; whilst the tonic effect on the nervous and muscular systems produces a desire for physical exertion, and makes work feel lighter.

Hydro-therapeutic reaction for heat

In hot water[1] treatment the most marked phenomenon is the dilatation of the superficial blood-vessels, which passes slowly off when the application is discontinued. With the dilatation of the surface vessels are associated increased secretion of sweat, and greater frequency of respiration. This triple effect of the treatment constitutes the reaction of the body to heat, and the three phenomena must be interpreted as the means by which the animal mechanism produces an increased loss of heat to counteract the heating effect of the hot application. The sweating is of course greatest in a hot air bath, though more visible in a hot vapour one.

Result of hot water applications

When the application of heat is general and sufficiently prolonged, a decided sedative effect follows the preliminary excitation, and is probably due in part to a certain anæmia of the brain and internal organs accompanying the superficial vaso-dilatation, and in part to diminished combustion in the tissues, accompanying diminution in the amount of heat production required to maintain the body temperature. A diminished desire for exercise is part of the general sedative action following hot baths. The somewhat constipating action on the bowels may be due in part to diminished peristalsis,

[1] Hot baths are those of or over 96° F.; very hot baths are those of 104° to 114° F., and any higher temperature that can be borne. Baths are tepid at 85° to 95° F. (ordinary temperature 90° F.). Any bath below 70° F. is called cold.

and in part to a diminution of intestinal secretion contrasting with the increased secretion of sweat.

Local applications

When the application of cold or hot water is limited to one part of the body instead of being general, certain remote reactions have been observed to take place, forming an additional argument in favour of the phenomena of the general reaction being of reflex nervous nature. Apparently when a limb is immersed in cold water, the corresponding limb of the opposite side reacts with it, becomes like it colder, and like it shows diminution in volume as measured by the plethysmograph, the diminution in volume being doubtless due to reflex vaso-constriction. Inverse phenomena take place at another level of the body; thus, during the application of a cold hip-bath, Winternitz observed an increase in the volume of the arm.

Utility of hydro-therapeutic treatment

Simple hydrotherapeutic treatment is often combined with residence at some health resort, and in such cases, change of air, food, occupation and mode of life must contribute largely to the result arrived at. Hot water, air or vapour baths (local or general), and the dry or wet pack, may be useful in 'muscular rheumatism' and lumbago, and in some cases of sciatica and various neuralgias; treatment by diet, massage, &c., is often combined. In chronic rheumatism, gout, and the uric acid diathesis, the milder hydrotherapeutic procedures are usually preferred, combined with the regulation of diet and exercise, and the diuretic effect of water taken internally. Douches, combined with massage, are often employed for the stiffness and thickening produced by chronic rheumatism, gout, and old injuries about the joints.

Cold water treatment may be useful in cases where it is desired to stimulate the general nutrition, as in slighter forms of anæmia and other cachectic conditions, and in some functional nervous disorders. It serves to 'harden the skin,' that is to render it less susceptible to reflex influences, and thus may be of use to persons with a great liability to 'catch colds,' to suffer from muscular pains, or repeated attacks of diarrhœa. It may be employed in the tonic treatment of convalescents, or in the 'after-cure' of persons treated by saline mineral waters for gastric catarrh, &c. Some patients with habitual constipation, and women with profuse menstruation of constitutional origin, derive benefit from cold water treatment. It may often be combined with massage or other treatment.

Contra-indications

For the success of cold-water treatment it is necessary that the organism can stand some abstraction of heat, that it can react to the stimulus of the cold, and that the digestive and assimilative organs are in fair condition. Especial care is needed in debility due to disease, and in the weakness of childhood and old age. Cold-water treatment is to be avoided in cases of chronic nephritis, considerable arterio-sclerosis, in all cases of aneurism, in tendency to hæmorrhage from the lungs or stomach, and where there has been an attack of cerebral hæmorrhage, or where the occurrence of an attack is feared. It is likewise contra-indicated in all cases of heart disease except slight, well-compensated affections of the mitral valve. In every case hydrotherapeutic treatment should be carried out under the supervision of a medical man.

Localities for hydrotherapeutic treatment

There exist a large number of establishments where hydrotherapeutic treatment can be obtained. It will be sufficient merely to enumerate some of them. In England there are those of Malvern, Matlock, Sidmouth, Conishead Priory near Ulverston, Ben Rhydding,

Ilkley, etc. In Scotland there are those of Dunblane, Crieff, Peebles, Wemyss Bay, Pitlochry, and others. In Ireland there is St. Anne's Hill (Blarney in County Cork). In Germany and Austria there are those of Nassau on the Lahn; Godesberg on the Rhine, near Bonn; Marienberg and Mühlbad, at Boppard on the Rhine; Laubbach, near Coblenz; Ilmenau, Liebenstein, Sonneberg, Elgersburg, and Schleusingen, in the Thuringian Forest; Bad Nerothal and Dietenmühle at Wiesbaden; Reinbeck (Sophienbad), near Hamburg; Teinach, in the Würtemburg Black Forest; Lauterberg, in the Harz mountains; Wilhelmshöhe, near Kassel; Schweizermühle and Koenigsbrunn, in the 'Saxon Switzerland'; Koenigstein, in the Taunus; Alexandersbad, near Wunsiedel, in Bavaria; Graefenberg-Freiwaldau, in Austrian Silesia; Kaltenleutgeben, not far from Vienna; Kaltenbrunn, near Voeslau; Kaltenbach, at Ischl in Austria; and very many others. In Switzerland there are the establishments of Champel, near Geneva; Aigle-les-Bains; Rigi-Kaltbad; Schönbrunn, near Zug; Schöneck, above the Lake of Lucerne, and other places. In France there are establishments at Paris, Auteuil, Gérardmer, Divonne, Bordeaux, Nice, etc. There are likewise establishments for hydrotherapeutic treatment at most of the chief Continental spas. Those in England are fewer in number, and usually more of the nature of ordinary hotels than those abroad.

CHAPTER II

CONSTITUENTS AND CLASSIFICATION OF MINERAL WATERS

Definition of natural mineral waters

NATURAL mineral waters form a part of Materia Medica, and from very early times have been employed in the treatment of disease, either internally or in the form of baths. As a convenient definition, we choose the following: Natural mineral waters are all waters which, as obtained from nature, are distinguished from ordinary waters either by the salts or gases they contain in solution or by their temperature being elevated.

It may seem a stretch of the term to include as mineral-water springs simple thermal ones hardly differing from ordinary springs except by the elevated temperature of their waters, but it is certainly convenient, as well as the custom, to do so. It must, moreover, be remembered that before chemical analyses were made attention had been drawn to many springs on account of the natural warmth of their waters, rather than on account of any special taste or smell due to peculiar chemical constituents. The Roman remains at thermal springs, such as Bath in England, bear abundant evidence to this. In other cases probably the smell of hydrosulphuric gas, the taste of Epsom salt, common salt, or iron, or the appearance of ochreous or other deposits, first drew special attention to the springs, though in many cases some curious tradition, superstitious belief, or ceremony was, later on, connected with the origin of the use of medicinal springs.

Not all mineral waters are suitable for use in medicine. Some of the strong iron waters, such as that of Sandrock in the Isle of Wight, contain too much of the irritant sulphate of iron to be used internally in ordinary cases of anæmia. Some waters contain too much of the sulphate and carbonate of lime. The sea, one of the strongest of mineral waters, though of great use for bathing, is seldom employed internally on account of the excess of common salt and disagreeable taste.

Method of classification

Various classifications of mineral waters may be attempted. They may be arranged according to their natural temperatures, according to their chemical constituents, or according to their therapeutic action. All classifications have their disadvantages, but the division into groups, according to the chief active chemical constituents of the different waters, is found most convenient, and is adopted in some form or other in nearly all works on mineral waters. Some waters are chiefly employed externally, others internally, but this will be referred to when the individual spas are considered.

Constituents of mineral waters

The list of elements which have been recognised—at all events, in traces—as present in mineral waters is very large, but it will be here more practical to enumerate the chief chemical combinations in which the elements occur dissolved in the waters. These are chloride of sodium (common salt), sulphate of sodium (Glauber's salt), sulphate of magnesium (Epsom salt), carbonate of sodium, sulphate of calcium (gypsum), and carbonate of calcium (chalk). In small quantities there occur carbonate of iron (the protocarbonate or ferrous carbonate), sulphate of iron (both the protosulphate or ferrous sulphate and the persulphate or ferric sulphate), chloride of iron (the protochloride or ferrous chloride), and crenate of iron, the bromides and iodides of sodium, magnesium and potassium, sulphides of sodium and calcium, arseniates of sodium, calcium, magnesium and iron, sulphate of copper, etc.

Together with the chlorides, carbonates, and sulphates previously mentioned are found the chlorides of calcium, magnesium, barium, strontium, lithium, potassium, ammonium, and manganesium, the carbonates of magnesium, potassium, lithium, strontium, manganesium, etc., sulphates of potassium, aluminium, and manganesium, borates, nitrates, phosphates, and silicates; the occurrence of these salts in minute quantities or traces is generally of little therapeutic interest, but the chlorides of calcium, magnesium, and barium probably occur in sufficiently large quantities to exert some effect (see pp. 25 and 26). Other mineral substances are found in traces, including the rare metals cæsium and rubidium, which were first detected in the mineral waters of Dürkheim.

Gases present in mineral waters

The most important gases dissolved in mineral waters are carbonic acid gas and sulphuretted hydrogen. Some waters contain an unusual amount of oxygen and nitrogen. The inflammable carburetted hydrogen, or marsh gas, is occasionally found (Porretta and Acireale, etc.), and in the waters of Harkany, in Hungary, the inflammable gas, carbonyl sulphide, or oxysulphide of carbon (COS), was discovered in the year 1867 by Karl von Than. In 1895 the chemically indifferent gas 'argon,' shortly after its discovery (1894), by Lord Rayleigh, as a constituent of atmospheric air, was found to be present in the thermal waters of Bath, Buxton, Wildbad, etc. The gas 'helium,' previously to its discovery in certain minerals in 1895 by Professor Ramsay, had been only known to exist by its band in the spectrum of the solar chromosphere, which was first discovered by Norman Lockyer and E. Frankland during the sun's eclipse of 1868; shortly after Professor Ramsay's discovery it was found to be present in association with 'argon' in the waters of Bath and of one of the Cauterets springs; though, according to Lord Rayleigh, probably hardly any traces of it exist in the earth's atmosphere. In some waters there are likewise organic and living organised substances, such as the jellylike barégine, which derives its name from being present in the waters of Barèges.

The classification adopted is that into:

Simple thermal group of mineral waters

1. *Simple or indifferent thermal waters* ('akratothermal waters,' or 'eaux oligo-métalliques chaudes').—These are poor in solid and gaseous substances, of low specific gravity, almost tasteless, of great transparency and softness. Their temperature lies generally between 80° F. and 150° F. Some contain an unusual amount of oxygen, some of nitrogen, but this has probably no special therapeutic significance. On account of their being frequently situated in wild mountainous regions the name 'Wildbäder' (*Thermæ silvestres*) has been given to this class of baths. There is no proof that the electrical conditions of these waters are peculiar, as has been suggested.

Muriated group

2. *Common salt or muriated waters.*—The first name is derived from their chief solid constituent—common salt—which, however, is likewise present in some waters of the other groups. The second name, 'Muriated Waters,'[1] is preferable on account of the presence, sometimes in appreciable amount, of other chlorides (the

[1] It would, perhaps, be more strictly correct to keep to the older nomenclature, and call these waters 'Muriated Saline Waters,' that is,

chlorides of calcium, barium, magnesium, lithium, potassium, and strontium). In addition to chlorides, small quantities of bromides and iodides (especially those of sodium and magnesium) and other salts are occasionally present; these admixtures may modify the action of the common salt. Many waters of this group are rich in free carbonic acid gas.

Alkaline groups

3. *Alkaline waters.*—In these carbonate of sodium is the most prominent constituent. They likewise contain carbonic acid gas, and according to the presence or absence of much chloride of sodium or sulphate of sodium may be subdivided into three classes:

(*a*) Simple alkaline waters.
(*b*) Muriated alkaline waters.
(*c*) Sulphated alkaline waters.

Sulphated groups

4. *Sulphated and muriated sulphated waters.*—In these sulphate of magnesium and sulphate of sodium, with or without common salt, are the chief ingredients. Those tasting of the bitter magnesium are commonly termed 'bitter waters.' Most of this group are used as aperient waters at home, but some of the 'muriated sulphated' division, containing comparatively much common salt, are used at the spas themselves (Brides-les-Bains, Cheltenham, Leamington, etc.).

Chalybeate group

5. *Iron or chalybeate waters.*—This group includes

mineral waters, the *salts* dissolved in which are chlorides (*muriates*). On the other hand, the term 'muriated' has the advantage of being shorter than 'muriated saline,' and a shorter term is preferable, especially when speaking of compound mineral waters, such as muriated chalybeate, muriated alkaline, etc. Moreover, the term 'saline' is often used both in England and Germany in special reference to the purgative salts, sulphate of sodium and sulphate of magnesium, and thus 'muriated saline waters' might be wrongly understood to mean the waters which we have here called 'muriated sulphated.'

those waters in which iron is contained in sufficient amount to confer on them a therapeutic action. The iron in mineral waters is usually contained in the therapeutically more valuable form of the bicarbonate, more rarely it occurs as protosulphate, persulphate, or protochloride. Occasionally it is associated with the presence of arsenic. Alum is sometimes present in sulphate of iron waters. Chloride of iron is associated with barium chloride and calcium chloride in the chloride of iron spring at Harrogate.

Arsenical group

6. *Arsenical waters.*—Waters which contain a sufficient amount of arsenic to exert a special therapeutic effect are conveniently classed in a group by themselves. In some waters, however, as those of Mont Dore, which have for convenience been placed in this group, it may be doubted whether the arsenic is present in sufficient quantity to exert any therapeutic action. In the strongest waters of this group the arsenic accompanies sulphate of iron; it likewise occurs in association with bicarbonate of iron, chloride and bicarbonate of sodium, etc.

Sulphurous group

7. *Sulphur waters.*—These contain sulphuretted hydrogen or a sulphide of sodium,[1] calcium, potassium, or magnesium, in appreciable amount. Some are thermal, others cold. Some are simple, others are compound, containing an admixture of common salt or other salts, sufficient in amount to exercise an influence on their therapeutic action. The total of solids found in solution in sulphur waters is, however, usually very small, and especially is this the case in the sulphide of sodium group, of which the Pyrenean sulphur spas (Bagnères-de-Luchon, Amélie-les-Bains, etc.) may be considered the representatives.

[1] This is usually present as a monosulphide, but in some waters also as a polysulphide (*e.g.* Barèges).

Low forms of living organisms flourish in the thermal sulphur waters, especially those belonging to the vegetable world (such as *Byssus lanuginosa*, etc.), and these give rise to the flaky, jelly-like substances, glairine, barégine, etc., usually found in this class of waters. Calcium sulphide, when occurring in waters containing sulphate of calcium, is supposed by Egasse and Guyenot to be sometimes due to the passage of these waters through soil rich in organic material, which, by withdrawing the oxygen from the sulphate, they suppose gives rise to the sulphide. Waters in marshy districts may contain sulphuretted hydrogen, resulting from the decomposition of vegetable matter; such waters may be suspected of conveying the germs of malaria, where this disease exists.

Earthy group

8. *Earthy or calcareous waters.*—Here the chief constituents are carbonate and sulphate of calcium, and carbonate of magnesium. These waters may be termed 'alkaline earthy waters' when carbonates of calcium and magnesium are the chief constituents, and may be termed 'gypsum waters' when sulphate of calcium forms the chief constituent. Many earthy waters contain varying amounts of iron, sulphur, common salts, etc., sometimes making it necessary to classify them also in other groups. Waters containing very minute amounts of both carbonate of sodium and carbonate of calcium are more conveniently considered as very weakly mineralised members of the simple alkaline than of the earthy alkaline group, but this relates to the subject of 'table waters,' on which there is a special chapter.

Table waters, &c.

9. *Table waters and other very weakly mineralised cold waters.*—Cold waters belonging to one or other of the preceding groups, but very weakly mineralised, often form pleasant 'table waters' on account of the carbonic

acid gas they contain, and are classed as a separate group, analogous to the simple thermal group, but cold and gaseous. To this group may be conveniently added some other weakly mineralised cold waters (French *eaux oligo-métalliques froides*), which can hardly be classed in one of the previous groups, but still deserve notice on account of a special therapeutic influence attributed to them. Such waters are those of Krankenheil, with a total mineralisation of about 1 per mille (containing a minute quantity of iodide of sodium), and those of Saint-Christau in the French Pyrenees, interesting for the minute quantity of sulphate of copper they contain, though their total mineralisation is only about ·3 per mille.

Defects in classification.

No classification of natural mineral waters, however elaborate, can be really perfect; for, in their constituents and the relative proportion of the constituents, they present infinite varieties. The classification which has been here adopted is, however, believed to be that which is practically found to be most convenient for reference and memory. Several of the spas should be mentioned under two or three different groups, either on account of the same spa possessing mineral springs which belong to different groups, or on account of a single spring in the character of its active chemical constituents falling between two groups.

CHAPTER III

ACTION OF MINERAL WATERS ON THE BODY IN THEIR EXTERNAL AND THEIR INTERNAL EMPLOYMENT

EXTERNAL USE

Mineral waters when not rich in salts and gases, if applied in the form of baths (this term being used in its broadest sense to include the various forms of douches, &c.), exert probably nearly the same effect as simple water would when applied in the same manner and at the same temperature. The effects of the external application of simple water have been already discussed (pp. 3–8). Most of the natural baths are either naturally warm, or are artificially warmed for use; warm and tepid baths, especially when alkaline, macerate the epidermis and have a greater cleansing effect on the skin than cold baths; they thus promote the excretion of the cutaneous glands. This diaphoretic effect is, of course, much increased in very hot baths and in natural vapour baths. (Monsummano and Battaglia.) Very many mineral water baths are given at tepid temperatures; such a bath constitutes a medium of uniform temperature enveloping the body, and acts in part by its soothing effect on the peripheral nerve-endings in the skin.

In the course of a bath, it has been found that

Absorption from baths

hardly any water is absorbed through the skin. Increased diuresis, when it follows baths, must be caused by reflex vaso-motor effects due to stimulation of the cutaneous nerve-endings, although at one time it was taken as evidence that water had been absorbed into the circulation from the body. Salts dissolved in the bath were likewise supposed at one time to be absorbed through the skin, but experiments have failed to prove this. Doubtless they may pass through any portion of mucous membrane with which they come in contact, but not through the healthy skin.[1] Salts dissolved in the bath-water may, however, saturate the epidermis, and by coming in contact with the outermost nerve-endings, impart a stimulating effect to the bath. The greater stimulating effect of brine baths ('Soolbäder') over baths of plain water is, of course, generally admitted.

Effect of gases in bath-water

Gases dissolved in a bath may pass into the circulation, as has been proved in the case of sulphuretted hydrogen gas. That the sulphuretted hydrogen absorbed in this way from the water of sulphur springs is sufficient in quantity to have any therapeutic effect is very unlikely. Still less is it likely that the free carbonic acid gas, in which some baths are rich, can exert any action by absorption through the skin, for owing to the pressure of the CO_2 already present in the blood, very little more is likely to enter from the bath-water through the skin. Some of it may be inhaled as it escapes from the bath, but the special stimulating effect of 'iron

[1] The absorption of drugs when applied to the healthy skin in the form of ointments or oils cannot be adduced as evidence against this, for these are rubbed or pressed into the skin. Mercury is so easily volatilised that when several patients are being treated by mercury baths or inunctions other patients in the same room may present signs of mercurialism.

baths,' such as those of Schwalbach, etc., is probably chiefly due to a mechanically stimulating action exercised by the bubbles of carbonic acid gas, as they collect and move along the skin. The stimulating effect of the warm salt baths of Nauheim and Oeynhausen ('Thermalsoolbäder') is partly due to a similar action of the gas, and in recent times it has become possible to fairly well imitate this by artificially charging salt water baths with carbonic acid gas.

To explain the action of the Nauheim baths (and of warmer baths[1] more weakly mineralised), in cardiac affections, it has been suggested that the vagus nerve is reflexly stimulated, and that thus a tonic effect is produced on the heart, giving rise to a temporary alteration in its movements, and that this may have a beneficial influence on the cardiac nutrition, comparable to that which mild gymnastic exercise brings about in the case of voluntary muscles. The ultimate good effect of a course of these baths is probably largely due to an increased power of excreting the waste products of metabolism, which a deficient circulation has allowed to accumulate in the blood and tissues of the body.

Gas baths

It may here be mentioned that gas baths of carbonic acid have been, and are, employed at some spas, the gas obtained from the mineral water being made use of for the purpose. The patient sits thinly clothed in an atmosphere of the gas, but either by a partition around the neck or by a properly placed overflow pipe care is taken

[1] In this connection it may be remembered that Dufresse de Chassaigne wrote in 1859 on the effects of the hot baths of Bagnols in chronic cardiac affections. The advantage which can be claimed for the Nauheim baths, is that, owing to the stimulating effect of the carbonic acid gas and salts, they may be administered at a lower temperature than baths of other mineral waters.

that none of the gas be inhaled; the gas has likewise been employed as a local bath or douche to various parts of the body. Similarly baths of sulphuretted hydrogen gas are sometimes employed. The utility of such gas baths remains doubtful.

Peat baths

Peat baths and mud baths are employed at a great number of spas on the continent. ('Moorbäder,' 'bains de tourbe,' 'Mineral-Moorbäder,' 'Eisen-Mineral-Moorbäder,' 'Schlammbäder,' 'bains de boue,' 'Mineralschlammbäder,' 'Schwefel-Moorschlammbäder,' etc.) The peats and peaty earths used for baths consist of decaying plant matter and soil; some of the peats are very rich in soluble salts, notably sulphate of iron, and contain free acids, such as sulphuric and formic acid. The disintegrated peat or turf of Franzensbad, when ready for use in the baths, is said to contain as much as 25 per cent. by weight of substances soluble in water, amongst which the large amount of sulphate of iron is considered especially important. The chalybeate peat of Marienbad is said to be even richer in iron.

Action of peat baths

These baths act as very large poultices to the surface of the body. Besides their thermal action, the effect of the weight on the cutaneous circulation may exercise some effect, and the sulphate of iron and the other salts and acids exercise a stimulating effect on the cutaneous nerve-endings. They are employed in chronic rheumatic affections, muscular pains, sciatica, local anæsthesia associated with sciatica, etc., remnants of inflammation in the pelvic organs, etc. The upper part of the thorax should as a rule not be immersed, and in diseases of the heart and lungs these baths are contra-indicated, for the weight of the bath pressing on the abdomen and lower part of the thorax may give rise to respiratory difficulties. After the bath is over complete rest for a short time is advisable.

Mud baths

For the mud baths a material is obtained from the mineral spring, consisting of precipitated salts, organic matter and material derived from the neighbouring soil. Their action is similar to that of the peat baths, but they are more fluid. The 'Schwefel-Moorschlammbäder' of Nenndorf, Meinberg, Wipfeld, etc., are made by the mixture of a sulphurous mineral water with a peat-like mud. Local peat baths and mud baths resemble the application of poultices to the diseased or painful part. At some places in Sweden a sort of massage with cold mud is employed. At Sandefjord in Norway a salt mud from the coast is applied like a hot poultice, or used for rubbing parts of the body.

Sand baths

At some health resorts local or general 'sand baths' are employed; the island of Ischia has been especially noted for their use. Baths in hot dry sand act doubtless somewhat similarly to hot air baths; they were used a long while ago, but have lately received proper scientific attention.

INTERNAL USE OF MINERAL WATERS

Internal use of simple thermal waters

The effect of drinking indifferent thermal waters probably resembles that of drinking an increased amount of ordinary pure water, and the therapeutic use of this has already been discussed in the chapter on Hydrotherapeutics (*vide* p. 2). It must, however, be remembered that the warmth of these springs makes some difference, for warm water naturally abstracts less heat from the body, and is more easily absorbed from the stomach than cold water. Warm water increases the action of the bowels less than cold water. The remarkable action described as being occasionally produced by drinking a single glass of some indifferent thermal water must be ascribed either to the imagination of an excitable patient, or to a temporary reflex effect on the circulation accompanying the mere act of drinking or sipping.

Common salt

Common salt is a normal constituent of the body. Moderate doses assist the digestion of albuminous substances, and stimulate the gastric mucous membrane, increasing the secretions of the stomach, intestine, and liver. The general nutrition is promoted, for sodium chloride, unlike the carbonate, seems to promote absorption and assimilation of nutritive material, as well as to accelerate downward tissue changes and excretion. Large doses (above five drachms, or even less, daily) may cause gastric irritation in some persons, and very large doses give rise to severe purgation and vomiting. Its power of local stimulation is much increased when given in concentrated solutions.

Carbon dioxide

Carbonic acid gas allays unpleasant sensations in the stomach, increases peristalsis and secretion, and thus furthers the effect of common salt waters. A large quantity, unless got rid of by eructation, may cause unpleasant symptoms either by distension of the stomach or absorption into the circulation.

Sodium carbonate

Carbonate and bicarbonate of sodium act as antacids, allay gastric irritation, and stimulate the flow of gastric juice. They alkalise the blood, and seem often to aid the action of iron in anæmia. They exert a diuretic action, and probably increase the action of simple water in 'washing out' the blood and tissues. Most of their good effects are obtained by repeated small doses, whilst large doses may cause depression, and, if long continued, emaciation. Their action is of course modified in those waters where they occur associated with common salt (see page 14), much carbonic acid gas, or the sulphates of sodium and magnesium.

Sulphates of sodium and magnesium

Sulphate of magnesium and sulphate of sodium, when they constitute almost the only active ingredients of springs, impart to them a merely laxative action. Such waters are mostly used at home. In the muriated sulphated and the alkaline sulphated waters the laxative quality of the sulphate is maintained, though the presence of the chloride and carbonate of sodium considerably modifies its action.

Iron

Iron, when present in sufficient amount, especially in the less irritating form of a carbonate, exerts its beneficial action on the quality of the blood in anæmia by increasing the number of red blood corpuscles and the amount of hæmoglobin they contain. This action is often favoured when the water likewise contains sodium carbonate and carbonic acid gas. Though only very little of the total amount of iron which has been swallowed is absorbed from the bowel, some is certainly absorbed,[1] and part of the effect of the portion absorbed on the blood may, it has been suggested, be due to a power of stimulating the hæmatopoietic function of the red bone marrow. C. Genth finds that gaseous chalybeate waters, like those of Schwalbach, exert a diuretic influence and cause an increased excretion of urea.

Manganesium salts

Minute quantities of the salts of manganesium occur in some mineral waters. The salts of this metal have been supposed by some to exercise a tonic action similar to that of iron, or to increase the efficacy of iron when

[1] Dr. A. B. Macallum, in the *Journal of Physiology*, vol. xvi. 1894, p. 268, showed that inorganic iron compounds are absorbed by the intestinal mucous membrane of guinea-pigs and other animals to an extent which varies with the nature of the compound and the quantity of it given.

given simultaneously with it. By others this action of manganesium is altogether denied.[1]

Arsenic

Arsenic is present in appreciable quantities in some waters, chiefly of the chalybeate class (see page 199), and may impart to them some of its beneficial influence in scrofula, in various kinds of malnutrition and anæmia, and possibly in psoriasis and skin affections.

Iodides and bromides

Iodides and bromides are present in the waters of Woodhall Spa, Hall in Upper Austria, Salzburg in Transylvania, the Adelheids-quelle of Heilbrunn, Castrocaro in Italy, Wildegg and Saxon in Switzerland, Salies de Béarn, Kreuznach, etc., but hardly in sufficient amount to make it certain that they exercise any therapeutic effect. In some 'Mutterlaugen,' of course, these salts occur in more considerable amounts; in that of Rothenfelde there is said to be 12·6 per mille bromide of magnesium.

Sulphur

The action of the sulpuretted hydrogen and minute quantities of the sulphides present in sulphur waters is not easily to be estimated. The effect of the weaker of these waters is probably due to other ingredients they contain, or is that of simple thermal waters. When used for a long period, however, the stronger sulphur waters have a tendency to produce anæmia (see p. 207).

Lithium salts

It is doubtful whether the lithium salts present in the waters of Baden Baden, Royat, etc., are taken in sufficient amount to produce any special therapeutic effect in gout, etc.

Calcium chloride

Chloride of calcium, said to be of use in scrofulous enlargements of glands, and in hæmophilic conditions,[2]

[1] See 'The Causes and Treatment of Chlorosis,' by Dr. Ralph Stockman, *Brit. Med. Journ.*, 1895, vol. ii. p. 1475.

[2] See 'On the Treatment of Hæmorrhages and Urticarias, which are associated with Deficient Blood Coagulability,' by Prof. A. E. Wright, of Netley.—*Lancet*, January 18, 1896.

is present in several muriated waters, and forms the chief part of the salts contained in the 'Mutterlauge' of Kreuznach.[1]

Barium chloride

In the Llangammarch Wells of Central Wales, chloride of barium is present, together with chloride of calcium. Barium chloride in very small doses is said to increase the strength of the heart's contraction whilst diminishing its frequency. Barium chloride and calcium chloride occur associated with iron in the chloride of iron spring at Harrogate, and with sulphur at the old sulphur wells of Harrogate.

Earthy salts

In the earthy and calcareous waters, the carbonate of lime has an antacid and soothing effect on the gastric mucous membrane, whilst the sulphate of lime is slightly astringent. This astringent quality need not always cause constipation, for, although the intestinal secretion may be lessened the peristalsis may remain the same, or even be increased. Usually, however, they exert a somewhat constipating action, and the diuretic effect of drinking the water is thereby increased, for when less fluid passes off by the bowel, more has to pass off by the kidneys; to this diuretic effect is probably due some of the repute of some of these waters in case of urinary gravel, etc. It is very doubtful if the lime in these waters has any special action on the nutrition of the bones, as has been suggested.

Empirical element in spa treatment

It is impossible to estimate exactly the effect of a given mineral water merely by summing up the respective effects of the ingredients which a chemical analysis shows it to contain. Empirical results have still largely to be relied on in treatment by mineral waters.

Inhalation of water

Inhalation Treatment.—Besides being taken by the

[1] In some 'Mutterlaugen,' such as those of Kissingen and Salzungen, there is less chloride of calcium, but more chloride of magnesium.

mouth the waters of some spas, especially those containing chloride and carbonate of sodium and sulphur, are inhaled for affections of the respiratory system, in order, if possible, to obtain the local action of the mineral water on the affected mucous membrane. When the pharynx or the naso-pharynx is the part affected, a coarse spray may be inhaled, so also when it is desired only that the spray should reach the upper part of the larynx. In cases, however, of chronic bronchitis, when it is desired that the spray should reach the mucous membrane of the bronchial tubes without exciting cough, it is necessary, especially when there is laryngeal irritability, to have the mineral water exceedingly finely pulverised. This is best effected by one of the methods which fill the entire room with the pulverised water. In such a room the patients can sit comfortably and inhale the spray which pervades the room.

Gradir-häuser

Another method of inhalation is that of sitting close to 'Gradirhäuser,' structures which were originally merely used for the manufacture of common salt from salt springs. The Gradir-häuser at Kreuznach, Kissingen, etc., are large fences of sticks, down which water from the spring is made to trickle. There are walks and seats arranged for the patients at the sides of some of these fences, and patients usually sit on the side away from the wind. Doubtless in addition to watery vapour and gases from the water, particles of the concentrated water itself are inhaled. A mineral water fountain in the neighbourhood will increase the amount of particles of water present in the air.

Inhalation of gases

The gases given off from mineral waters, especially sulphuretted hydrogen, nitrogen, and carbonic acid, are sometimes inhaled, but it is very doubtful if any real thereapeutic use has been obtained by this method.

Other factors in spa treatment

So far we have discussed the physical effect of baths and the pharmaco-dynamic action of mineral waters when taken internally. These effects of the mineral waters might often be obtained by their judicious employment at home, or, at all events to a large extent, by the employment of artificial mineral waters, but treatment at spas depends for its success also on other factors, and it is our intention in the following chapter to take into consideration the mental repose or change in mental occupation, the change in climate, surroundings, mode of life and diet, accompanying spa treatment.

CHAPTER IV

THE INFLUENCE OF CHANGE OF AIR AND DIET AND HABITS IN THEIR CONNECTION WITH SPA TREATMENT — ORDINARY MEDICAL AND SURGICAL TREATMENT AT SPAS—MUSCULAR EXERCISES AND MASSAGE IN CONNECTION WITH SPA TREATMENT

It is always very hard in estimating the effects of spa treatment, to separate that which is due to the mineral waters from that which is due to change of air, diet, mode of life, and mental occupation. If these latter did not contribute largely to the good results obtained, and if the results were merely due to the internal and external use of spa waters, it would be possible in most cases to carry out the treatment at home with the aid of imported or even artificial mineral waters. As a matter of fact the home treatment often fails, or its result falls far short of that obtained at the spas themselves. Nor is this to be wondered at when we consider the effect alone of 'change of air.'

Change of air, etc.

Every reader has probably experienced some of the effects attributed to 'change of air.' It would be quite unnecessary to describe, if that were possible, what all feel when from the confined air of rooms or offices in some large smoky town, they go for their holiday and inspire the fresh fragrant country air, or the invigorating sea or mountain breeze. During the holiday much time is usually spent in the open air ; sunlight, including the invisible 'chemical' rays, which probably penetrate more deeply than ordinary light rays, plays some part in the good effect, probably increasing the general nutrition of the body, as well as rendering the air more aseptic.

Climate and altitude of spas

The elevation of some baths above the sea-level gives them a real mountain climate. St. Moritz in the Upper Engadine is about 6,000 feet above the sea ; Panticosa in Spain, Loèche-les-Bains in Switzerland, Bormio in Italy, and some others, are situated at elevations of 4,000 feet and 5,500 feet. Many other baths lie at considerable elevations. Buxton in Derbyshire has an elevation of 1,000 feet, and its situation imparts a delightful freshness to the air. It will be here impossible to enter at further length into the pure climato-therapeutic influences of spas. In selecting the spa the season also when the patient requires treatment must be considered, a spa being selected at which the climate during that part of the year is at its best.

Psychical influences in spa treatment

The good effects of taking an ordinary holiday, often attributed to 'change of air,' are doubtless often in part psychical, and due to change in mental occupation. The dull routine of office work, and the excitement and worries of commercial enterprise or professional life, are alike laid aside. The mind is occupied in other ways. Often country life has something of the charm of novelty; to others it recalls pleasant recollections and re-awakens earlier interests. Rest and quiet are usually most keenly relished by those who have been hardest worked.

If this psychical action is needed on ordinary holidays, it is still more important during spa-treatment. Though in some cases it may do good, a course of powerfully-acting mineral waters cannot be recommended in most cases, whilst the patient has to attend to a worrying business or social fatigues. Sometimes the only way to ensure that a patient keeps away from business cares, is to insist on his going to a health resort at some distance for treatment.[1]

There should, however, be no *ennui* at spas, and this is often well guarded against at foreign spas, where bands and concerts in the open air are provided and social entertainments looked to. Patients at spas should be made as cheerful as possible, and their thoughts should be diverted from their ailments by healthy mental influences. Neglect of this factor in spa treatment has led to want of success at some spas, whilst the due recognition of its importance has contributed largely to the success of other spas. In the treatment of chronic diseases, the mind can and should be used as a most powerful agent. Change of occupation and amusement act as psychical stimulants; they doubtless promote the nutrition of nerve-cells in the cerebral cortex, and through improved nutrition of the brain cause improvement of nutrition and function in distant organs.

Regulation of diet and mode of life

Another advantage of treatment at a spa over treatment at home is a certain amount of routine in the treatment, not to be regarded as a thoughtless, 'mechanical' method of treating all patients. A patient finds it easier to change his diet and habits when others are doing the same thing; it becomes, indeed, almost necessary to do so. Thus, over-indulgence in food (and per-

[1] The journey, however, should not be made too fatiguing by continuous 'through' travelling. If the health resort be distant, the journey should be interrupted by rests; by not observing this point patients are likely to arrive at their destination in a state of fatigue which renders a considerable period of rest advisable before commencing spa treatment.

haps alcohol), hurried meals at irregular times, and late hours give place to early rising, regulated diet, regular meal-times, and going to bed early. Regulation of the diet is particularly difficult to effect at home, and what a large part this plays in the treatment of obesity and glycosuria at Marienbad, Karlsbad, etc., is generally acknowledged.

Enough has been said to show how many advantages treatment at the spas themselves has over treatment by mineral waters at home.

Ordinary medicinal treatment at spas

In some cases it may perhaps be objected that the good results obtained by spa treatment are not so much due to the waters as to ordinary medicinal treatment employed there. In this there is doubtless occasionally some truth. Thus the reputation of Aix-la-Chapelle in syphilis has been due, in great part, to the ordinary medical treatment employed there and to the attention paid to the subject by the local doctors; the reputation of Karlsbad in diabetes is partly due to similar causes. Certain spas owe some of their reputation in gynæcological affections to the skilful local treatment employed by the doctors, and Wildungen is famous for the surgical treatment of urinary diseases.

In other cases the reputation of the spa is due not so much to the water as to the energetic hydro-therapeutic measures, special exercises, massage, etc., employed there. This is the case, to some extent, at Aix-les-Bains for joint affections, and to similar causes Nauheim owes much of its recent celebrity in heart affections.

As a general rule, ordinary pharmaceutical treatment is employed as little as possible at spas, patients rightly or wrongly thinking they have had enough of drug-treatment before they are sent to spas, or having a special dislike to medicines. What has been stated in the last paragraphs serves merely to emphasize the fact that the knowledge, capability, and energy of the local medical men contribute largely to the success of spa treatment, and this factor must always be taken into consideration when selecting a spa for patients.

Importance of the local medical men

Active and passive exercises in connection with spa treatment.—Massage is now occasionally employed in most spas: local massage in the treatment of stiff joints, sciatica, lumbago, &c., and general massage in the treatment of patients in whom much voluntary muscular exercise is unsuitable or impossible. In the latter cases general massage is intended, to some extent at least, to supply the place of voluntary exercise.

Massage in spa treatment

Institutions for Swedish gymnastics have been established at many foreign spas, furnished with Dr. G. Zander's medico-mechanical appliances for passive movements and for voluntary muscular exercises with graduated resistance. Institutions of this kind are to be found at Aix-la-Chapelle, Wiesbaden, Baden-Baden, Wildbad, Karlsbad, Ragatz, etc. Here graduated movements may be made in order to exercise particular joints and particular sets of muscles.

Swedish gymnastics and Nauheim exercises

Swedish gymnastics without special mechanical appliances, according to the original system of P. H. Ling, are practised under strict medical superintendence at Homburg, Baden-Baden, and other localities. In the 'Schott' treatment for heart affections at Nauheim, a form of 'Widerstands-Gymnastik' (movements with resistance), under the direct superintendence of a doctor or skilled attendant, has been introduced. In the systems of Ling and Schott the resistance is supplied by the hand of the superintendent, whereas in Zander's system it is supplied by the weights and levers of his machines.

Walking and climbing exercise

Graduated voluntary exercise in the form of the 'Terrain-Cur' was some years since largely introduced into spa treatment, especially at German spas. Paths have been made on the hills and slopes around the spas, which involve a varying amount of up and down 'climbing' exercise to those who go along them. Maps of the different walks are made, and in selecting a series of walks for his patient, the doctor can regulate the time, the length, and the amount of climbing exercise for every day's walk. The arrangements for the 'Terrain-Cur,' as this is termed, were introduced after the writings of Professor M. J. Oertel, of Munich, in 1886, had directed attention to his method of treating chronic heart affections by gradual climbing exercise. ('Ueber Terrain-Curorte zur Behandlung von Kreislauf-Störungen,' Leipzig, 1886). The use of exercise in certain heart affections had already been advocated by Stokes, who, writing on the treatment of 'incipient fatty diseases of the heart' (Professor William Stokes, 'Diseases of the Heart and the Aorta,' Dublin, 1854, p. 357) commences thus:—'We must train the patient gradually but steadily to the giving up of all luxurious habits. He must adopt early hours, and pursue a system of graduated muscular exercise, etc.'

Uses of exercise and massage

For the use of muscular exercise in headache associated with habitual constipation, etc., we refer to the remarks in Chapter XVIII., Section 46. Its great usefulness in many cases, by furthering the oxidation of the downward products of tissue-metabolism, is undoubted, and can hardly be over-estimated. By regular exercise much may be done to prevent the premature degeneration of the tissues, those of the vascular system in particular, to which persons with an inherited arthritic tendency are specially predisposed. By means of exercise the voluntary muscles are enabled to make use of the sugar circulating in the blood in glycosuric cases. On the action of regular exercise in helping to hinder excessive development of fat in the body, we need not enter here.

A moderate amount of muscular exercise in those who are able to take it, helps the body to get rid of waste products, and promotes the healthy nutrition of all the tissues. When, owing to debility, obesity, stiff joints, or certain affections of the circulatory and respiratory systems, sufficient ordinary muscular exercise, such as walking, has become impracticable, massage or some modified form of exercise can often be practised with advantage.

It is probable that courses of the various active and passive exercises, and of hydro-therapeutic treatment, all help to get rid of the toxic materials and waste products which have accumulated in the body. This is effected either by oxidation within the body, or by elimination in the urinary and other excretions. It is therefore of great interest to hear from Dr. Blanc, of Aix-les-Bains, that Monsieur A. Ranglaret has recently proved by injections into rabbits that the specific toxicity[1] of the patient's urine is increased on his commencing a course of the 'douche-massage' treatment (see p. 215).

[1] It is not improbable, as suggested, that the group of symptoms, known as 'well-fever,' often appearing during spa treatment, is due to a temporary excess in the toxic materials circulating in the blood previously to their elimination. If this supposition be correct, 'well-fever' may be regarded as analogous to the pains and stiffness felt at the commencement of a walking tour, or after any unusual muscular exercise in persons who are 'out of training.' [Exercise in persons out of training perhaps gives rise to the familiar pains and stiffness, firstly by inducing a too sudden catabolism in the muscles, secondly by bringing into the circulation waste products which were previously stored up in the tissues.] Bearing also in mind the analogy between the temporary

Massage and exercises in affections of the heart

Dr. Lauder Brunton and Dr. Tunnicliffe have lately shown (*Journal of Physiology*, December 1894) that massage causes a diminution of peripheral resistance in the vessels of the kneaded muscles, and that hence, soon after the kneading, an increased flow of blood through the part takes place, together with a fall of the general blood pressure. During the massage the blood pressure may be slightly increased; but this slight increase is not likely to throw a great amount of extra work on the heart, such as occurs at the commencement of climbing exercise. Hence, in cases where the coronary arteries of the heart are diseased, and any attempt at climbing exercise causes attacks of angina pectoris, massage may be employed as a substitute to voluntary exercise without inducing such attacks. In other cardiac cases, where for various reasons the patient can only take a very limited amount of voluntary exercise, this deficient amount may be supplemented by massage.

When only very little voluntary exercise is possible, it is sometimes best that it be taken under skilled supervision. In the use of Dr. Zander's machines this supervision is supplied by the medical attendant, who prescribes the exercises or who is present in the room whilst they are performed; in the systems of Ling and Schott the supervision is supplied by the doctor or skilled attendant who furnishes the resistance to the movements.

Dr. Brunton (*Lancet*, October 12, 1895) has pointed out why voluntary muscular exercise, in some cardiac cases, can have an advantage over general massage.

pains and stiffness resulting from unwonted muscular exercise, and those often complained of by gouty, rheumatic, and anæmic persons, it is not astonishing that such patients should often complain of an increase in their pains shortly after commencing spa treatment.

During exercise the respiratory movements are increased, and thereby a kind of indirect massage is practised on the heart and large thoracic vessels. Hence the importance of assuring, where possible, a certain amount of voluntary exercise, even when massage serves the main purpose. Swedish gymnastics often best supply this exercise, because in these forms the amount can be easily regulated and the movements varied as required. It is in the treatment of affections of the heart that the 'Widerstands-Gymnastik' is at present exciting especial attention. (See Nauheim, p. 83.)

CHAPTER V

DAILY LIFE AT SPAS—DURATION OF THE CURE—NECESSITY FOR MEDICAL SUPERVISION—SEASONS FOR THE CURE—IMPORTANCE OF AN 'AFTER-CURE'

A FEW words may be said of the daily life of patients whilst treated at foreign spas. This must naturally vary according to the strength of the patient, his previous habits, the nature of his ailment, and the kind of mineral water he is taking; it must necessarily be largely regulated by the spa doctor, and depends somewhat on local customs of the particular spa. A certain amount of routine is, however, often useful, for patients find it easier to follow rules when others about them are doing the same. Doubtless rules for drinking the waters, bathing, and regulation of diet, had formerly become too stereotyped at some spas, and were followed too rigidly without sufficient regard to the special condition of individual patients (see under Karlsbad, p. 150). More recently these rules have been wisely modified or relaxed to suit the requirements of different constitutions.

Daily life at spas

At most foreign spas the patient's day begins early. He gets up at six or seven, drinks his water, chats and promenades, whilst a band, to the expenses of which the patients subscribe, enlivens the Kurplatz. Breakfast consists of coffee or tea and rolls, to which, especially in the case of English patients, a couple of eggs or a little ham, chicken, etc., is often added. This can be taken between seven and nine, according to the time at which the patient has started his day. It is the rule to take the waters on an empty stomach, but in the case of delicate patients a cup of milk or tea or coffee can be allowed on rising; in some cases the waters may be drunk in the patient's room. Occasionally it is found preferable to delay taking them until before the mid-day or evening meal.

In Germany the middle meal of the day is at about one o'clock, in France it is earlier. If baths are ordered, they may be taken in the morning early, after drinking the waters, or, if there is not time then, before the mid-day meal. When the spa is overcrowded and the bathing accommodation not great, the time for the bath must depend on priority, and this at some spas may, during the height of the season, cause the patient great inconvenience and even harm. Open-air concerts, promenades, and occasionally pleasant excursions into the surrounding country, help to fill up mornings and afternoons until the evening meal at five or six. Sometimes drinking the water a second time before the mid-day or evening meal is ordered; occasionally, especially when it is advisable that the patient should drink very little at a time, or when an unusually large daily amount is required, the water is taken three times daily. Chalybeate waters are sometimes taken with the meals, and if a large amount of free carbonic acid gas disguises the taste of iron, they form pleasant table drinks.

Duration of the cure

There is no fixed time for the 'cure' to last. Three to four weeks is probably about the average, but this treatment, like any other treatment, must vary according to the patient's condition and the ailment. It can sometimes only be settled by the medical man who is watching the progress of the case, and in many cases of chronic disease it is necessary to continue the course over six or eight weeks, or to have two courses in the same year separated by an interval of one or two months.

Importance of medical supervision at spas

Medical supervision is absolutely necessary. The patient's progress must be watched. He requires advice on many points, as to his diet, as to the nature and amount of exercise, and the time of day at which he is to take it, as to when he is to drink the waters, and how much he is to take. Sometimes the doctor orders that the waters, if cold, be warmed; or, if too strong, be diluted with ordinary drinking water or milk or whey; or that their taste be improved by the addition of some gaseous water, etc. Many patients cannot be satisfactorily treated unless they have very precise rules to follow, and this exactness can only be furnished by a doctor at the spa itself. On his guidance the result of the treatment often depends, and he should receive from the medical attendant at home an account of the patient's condition and previous treatment. By his successful general management of cases the good reputation of a spa is often largely increased.

Careless use of mineral waters

Although very large quantities of mineral waters have sometimes been drunk with impunity or apparent benefit, even the indifferent waters should not be taken without caution and supervision. Very serious symptoms, and even death by syncope or apoplexy, have been known to follow the sudden drinking of cold water or excessive quantities of hot water. A lesser evil of drinking too large quantities is the disturbance of the whole

'Well-fever'

system known as 'well-fever,' 'Bad-Friesel,' or 'poussée'; it may also be produced by excess of bathing or other external use of the mineral water, and consists in febrile uneasy sensations, and sometimes dyspepsia, lassitude, diarrhœa, and skin eruptions. These symptoms, which were formerly supposed to be of a critical and beneficial nature, soon pass off with at the most a temporary cessation of the treatment or the administration of a sedative drug. At Loèche-les-Bains (Leukerbad) the eruption or 'poussée' is considered still a normal part of the treatment by prolonged tepid baths. 'Well-fever' may perhaps be compared to the pain and stiffness often felt even by healthy persons at the commencement of a walking tour (see p. 35).

Diet

The rules as to diet during the cure at some spas were formerly too strict and stereotyped; the same 'cure-diet' was observed no matter what the patient suffered from, the 'Sprudel-Suppe' supper of Karlsbad (see p. 150) being a favourite example of this severe dieting. Articles such as butter and tea were, without sufficient reason, prohibited in all cases. These rules have been somewhat relaxed, but it is still most important that the spa physician should be able to supervise the patient's diet. Table d'hôte dinners in this respect are somewhat inconvenient, and separate meals as at Karlsbad are often to be preferred. At some places, as at Karlsbad and Wildungen, the physician has considerable control over the sort of food which the spa guests receive.

Seasons for spa treatment

The season for spa treatment is necessarily limited to the time during which the spa is open. This is mostly from May till October, although some are only open from June to September. Bath in England is open all the year round, and some foreign spas, such as Aix-la-Chapelle, Amélie-les-Bains, Dax, Baden-Baden, and Wiesbaden, are likewise open throughout the year. The summer months are especially convenient for a cure, because the patient can remain in the open air; moreover, on account of the warmth of the air, less heat production and tissue metabolism are required, and therefore a better opportunity for the alterative and depletive action of mineral waters is afforded.[1] If a winter course be adopted, the patient should, if possible, have his lodging in the building in which his bath is, so as to render him independent of inclement weather; such an arrangement is possible at Bath, Aix-la-Chapelle, Wiesbaden, etc. For those who bear heat badly, it is advisable to avoid the hottest summer months at Aix-les-Bains, Aix-la-Chapelle, Ems, Baden-Baden, Wiesbaden, Neuenahr, Ragatz, and other hot localities.

Preparatory treatment

At one time preparatory treatment of a severe nature was advised before taking a course of waters. This was according to the antiphlogistic theories of the time. Preparatory treatment is still sometimes adopted, such as rest at some climatic health resort or special medicinal treatment, but not the excessive purging, etc., of olden days.

Use of the water after leaving the spa

Sometimes, on discontinuing a laxative course at spas, such as Karlsbad or Marienbad, there may be troublesome constipation. This may be remedied by continuing the use of the mineral waters, or their salts, for some time after leaving the spa.

Importance of of an 'after-cure'

Generally speaking, an 'after cure' is of the greatest importance, especially after the more active waters, such as Karlsbad, Marienbad, and Kissingen. Instead of going immediately to their homes and beginning their

[1] It is possible that in ordinary pharmaco-dynamic treatment feeble patients are better able to undergo courses of drugs, such as mercury, large doses of iodides, and thyroid preparations (which make demands on the patient's strength) in warm weather than in cold weather, when more of their energy is used up in heat-production.

usual mode of life again, patients should abstain from active work and keep to a simple diet and open-air life for some weeks. They may go to some pretty part of the country not far removed from the spa, or to some not very distant mountain health resort. For some time subsequent to courses of active laxative waters, the nervous system and bodily functions are in a specially sensitive condition, and are easily thrown out of order, as they are during convalescence from an infectious disease, by nervous excitement, business worry, or bodily fatigue.

During the 'cure' the patient gets rid of the unhealthy and effete material accumulated in his tissues. During the 'after-cure' a vigorous building-up process ought to take place, just as it does during convalescence from a disease, and new healthy material is assimilated by the tissues in place of the unhealthy material cast off during the 'cure.' Roughly speaking, something of this sort is what takes place, and neglect of the 'after-cure' may lead to disagreeable consequences, another breakdown, and the patient may lose all the good results of the treatment. At some spas the importance of the 'after-cure' seems still to be hardly sufficiently recognised.

Localities for the 'after-cure'

The nature and situation of the climatic health resort to be selected for rest, after the various courses of waters, is not without importance; but it is difficult to lay down general rules, as every case must be considered according to its individual nature and the accompanying circumstances. Courses of the more active waters usually ought to be followed by a longer rest than courses of the less active waters; but in the patients themselves there are great differences, which must guide the medical man in deciding the length of the after-cure to be recommended, and the particular health resort to be selected.

Some patients are so weak at the commencement of the treatment that even after a very slight course of waters a long rest is necessary, and the locality selected must not be too distant from the spa.

Localities of considerable elevation

If we have to deal with patients possessing a soundly-acting heart, we need not fear a somewhat longer journey, and are able to recommend bracing localities of considerable elevation (3,500 to 7,000 feet), even in the absence of level walks. Such localities are, in the northern and central part of Switzerland: the hotels on the Rigi and the Pilatus, Andermatt, Hospenthal, Disentis, Mürren, Wengen, Gurnigel, Lenk, and Grindelwald; in the northern and north-western part of the Grisons: Arosa, Klosters, Davos Platz and Davos Dörfli, Clavadel, Frauenkirch, Wiesen, Churwalden, Parpan; in the Upper Engadine: Samaden, Pontresina, St. Moritz, Campfer, Silvaplana, Maloja, and Zuz; on the heights of the Rhone Valley: Berisal, the Eggischhorn, the Rieder Alp and the Bel-Alp, Villars, Hôtel des Diablerets, Hôtel de Chamossaire, La Comballaz, Château d'Oex and Gryon; in the Swiss Jura: the Weissenstein; in the Monte Rosa district: Zermatt, the Riffel-Alp, Saas-Fee, Evolena, Arolla, Vissoye, Zinal; and on the Italian side of the Monte Rosa group: Macugnaga, Alagna, Gressoney St. Jean, and Hôtel Monte Generoso; on the Italian side of Mont Blanc: Courmayeur and Ceresole Reale; and on the northern side of Mont Blanc: Montanvert above the Mer-de-Glace near Chamonix; in the southern Tyrol: Campiglio, Cortina di Ampezzo, Schluderbach, and San Martino di Castrozza; in the Ortler district of the Tyrol: Sulden and Trafoi; Eggerhof above Meran, and Mendelhof above Botzen; in central Tyrol: Innichen and Alt-Prags; and in the Upper Valtellina: Bormio and the neighbouring Santa Catarina.

Localities of moderate elevation

If the patient's heart is dilated and feeble, high altitudes and the absence of level ground must be avoided, whilst moderate elevations (600 feet to 3,000 feet), with the opportunity of exercise on level or only gently rising ground, are preferable to low situations and even to the seaside. Such are many of the health resorts in the Black Forest: Badenweiler, Rippoldsau, St. Blasien, Griessbach, Petersthal, Allerheiligen, Freudenstadt, Titisee, Wildbad, Herrenalb, and Teinach; in the Thuringian Forest: Friedrichroda, Tabarz, Liebenstein, Ruhla, Oberhof, Ilmenau, Elgersburg, the Thuringian Blankenburg, etc.; in the Hartz mountains: Harzburg, Wernigerode, Ilsenburg, Gernrode, Alexisbad, Blankenburg, Ballenstedt, Clausthal, Andreasberg; in the Vosges mountains: Hohwald, Gérardmer; in the Fichtelgebirge: Alexandersbad, Berneck; in the Taunus: Kœnigstein, Schlangenbad, Schwalbach, Homburg; in the 'Franconian Switzerland': Streitberg and Muggendorf.

To the north of the Bavarian highlands are some agreeable localities for this class of cases, such as Starnberg and Tegernsee, on the lakes of these names; in Silesia Schreiberhau may be mentioned; the Salzkammergut contains many useful and charming places: Salzburg, Gmunden, Ischl, Aussee and Alt-Aussee, St. Wolfgang, Hallstatt, Zell am See; the central and northern Tyrol is very rich in localities of this character, we mention only Innsbruck, the Achensee, Partenkirchen, Garmisch, Kainzenbad, and Bruneck.

In patients with dilated hearts the differences in the individual power are so great that in some an altitude above 1,200 feet is not well borne, while others feel perfectly well at 3,000 feet or more. In thoroughly well-compensated mitral affections high elevations are often quite as well borne as when the heart is perfectly normal.

Localities for malarial cases

When cases of malarial origin have undergone spa treatment, this treatment should always be followed by a long stay at some locality perfectly free from malaria. Places of high elevation exercise a much better effect than localities at low elevations, especially those in close proximity to large glaciers, such as Montanvert, the Bel-Alp, the Rieder-Furka, the Eggischhorn, Pontresina, and Arolla.

For chronic rheumatic cases

A very important class of cases comprises those of chronic rheumatism. If the heart in these cases is sound, one rarely need be afraid of a high altitude, though in many of them lower elevations are equally satisfactory, but it is essential to select dry and sunny localities. Such localities are: Les Avants, Glion, Caux above Glion, St. Beatenberg above the Lake of Thun, Gurnigel, Pontresina, Maloja, Rigi-First and Rigi Kaltbad, and the localities on the heights above the Rhone valley, which have been previously mentioned. In lower positions, Badenweiler and Homburg are suitable.

For many rheumatic cases a long stay at the seaside, after a course of waters, is preferable to a stay at mountain health resorts, owing to the strengthening influences which the sea air exercises on the skin. Bathing in the open sea is, however, mostly to be avoided after the use of active spas, such as Karlsbad, Marienbad, Tarasp, Franzensbad, and Kissingen.

In cases of emphysema and bronchitis

In sufferers from emphysema and chronic bronchitis only moderate altitudes are borne, and it is desirable to select localities without wind and dust, situated if possible within, or in the neighbourhood of large forests, pine forests by preference. Such localities are the Flimser Waldhäuser (too high in very advanced cases), and Ragatz in Switzerland; Alt-Aussee (an hour from Aussee), Kreuth and Achensee, and Zell am See, in the eastern Alps; the localities in the Harz Mountains mentioned above; Badenweiler, Baden-Baden, Wildbad, Teinach, Griessbach, Rippoldsau, &c. in the Black Forest; Hohwald in the Vosges Mountains; Alexandersbad in the Fichtelgebirge; Friedrichroda and Liebenstein in the Thuringian Forest; Schlangenbad and Kœnigstein in the Taunus; Brückenau in Franconia.

Localities in the United Kingdom

It is sometimes important that the after-cure be made nearer home, and there are a number of localities in England, Scotland, Wales and Ireland suitable for residence after spa treatment, though they have not the high elevation which might be desired for some cases. Many of them find a place in our list of marine spas (pp. 281–283); others, such as Buxton, Harrogate, Llandrindod, Malvern, Tunbridge Wells, Strathpeffer and Bridge-of-Allan, are included amongst the ordinary spas; and others again, such as Ilkley, Ben Rhydding, Pitlochry, etc., are mentioned amongst the localities for hydrotherapeutic treatment (see pp. 8 and 9). There are also numerous inland places of the United Kingdom, not mentioned under one of these heads, which can be used for an after-cure. It is, however, scarcely necessary to mention them here, as suitable localities of this class are probably well known to the members of the medical profession in England.

If we were to enter fully into the climatic conditions of the localities to be recommended for different invalids, after courses of waters, we should go beyond the sphere of this book; but enough has been said, we suppose, to serve for guidance on this rather important matter.

Precautions during the 'after-cure'

Whatever locality may be recommended, the invalid must always bear in mind that it is essential for him to spend as much time as possible in the open air, to avoid fatigue, to continue with a strict diet, and to be very careful with regard to dress, so as not to expose himself to chills.

CHAPTER VI

SIMPLE OR INDIFFERENT THERMAL WATERS

These waters taken internally (see pp. 2 and 22) help in the removal of waste products from the tissues, and hence are useful in some cases of chronic gout and rheumatism, especially in those cases where more active waters are not advisable. By increasing the secretions and rendering the contents of the bowels more fluid, they may be of service in cases of constipation due to insufficient intestinal and biliary secretion. By their soothing local action, and by their indirect influence on the general nutrition, they may have a good effect in some forms of gastralgia and irritable conditions of the gastric and intestinal mucous membranes.

In the form of warm baths they exercise a sedative influence on the nervous system. Hence they may be useful in some cases of neuralgia, in hyperæsthesia, painful menstruation, nervous cough, and tendency to hysteria and irritable functional nervous affections.

As baths, these waters have also enjoyed a great reputation in the treatment of painful cicatrices (especially the hotter baths), and in the healing of troublesome wounds and ulcers.[1] In the latter class of cases they

[1] Thus Paracelsus, in his account of Pfaeffers (a genuine work of his), speaks of the curative action of these waters in cases of ulcers, sinuses, and incompletely healed wounds. In the Apothecaries Hall of London is a painting of the baths of Pfaeffers, and it may be presumed that

act doubtless, partly by improving the general health of the body, partly by a local action, similar to that of the prolonged local and general baths employed by surgeons in phlegmons, burns, etc. They clear away discharges from the surface of the wound, maintain an even temperature, and exercise a soothing effect on the exposed nerve-endings. In these days of antiseptic and aseptic surgery, this class of spa treatment will probably be less required.

The action of prolonged tepid baths as employed at Loèche-les-Bains in chronic cutaneous eruptions is probably somewhat similar. The tepid water macerates the epidermic scales of psoriasis, and washes away the scales and exudation of eczema, a soothing yet tonic effect on the nerve-endings being exercised by the continued application of the thermal water at a constant temperature. Some of the successful cases of former days may have been in persons afflicted with scabies, before the 'acarus scabiei' or 'sarcoptes hominis' had been discovered as the cause of the affection. In such cases the parasites may simply have been drowned in the long-continued baths. By their anti-parasitic action sulphur waters may have been still more successful than indifferent waters in such cases.

It seems difficult to understand how any permanent benefit can be derived from warm baths in organic affections of the nervous system; nevertheless, some indifferent thermal springs, such as those of Wildbad and Gastein, have obtained a reputation in the treatment of chronic spinal affections. It is certainly unlikely that warm baths judiciously used can do harm in commencing

Franc. Manning, British Minister to Graubünden, whose attempted assassination, June 27, 1711, is recorded on the picture, considered that he owed a debt to these waters.

cases, and by improving the general nutrition they may do at least temporary good in chronic spinal affections, and sometimes relieve the pains of tabes dorsalis. It is probable that in former times thermal baths sometimes got the credit of curing cases of paralysis which were really of functional nature, and cases of paraplegia and apparent tabes when these were really due to peripheral neuritis,[1] and in the ordinary course of events would have tended to recovery.

Much is now known of the ætiology of peripheral neuritis, and most cases can be traced to alcoholic drinks, lead, arsenic, and the toxines circulating in the blood after diphtheria, enteric fever, and other infectious diseases; but in other cases the cause cannot exactly be ascertained, and they are spoken of as rheumatic, idiopathic, &c. Some of the more chronic forms of neuritis, evidenced by pain, anæsthesia, paræsthesia, or even by loss of motor power in nerve areas, may possibly be caused by a vitiated condition of the blood, due to some one of the various possible kinds of auto-intoxication, and may be associated with a cachectic condition of the whole body. In such cases especially are thermal baths likely to do good by exerting a favourable influence on the general nutrition.

In chronic rheumatism and sciatica the hotter baths

[1] Some cases of peripheral neuritis, the so-called 'pseudo-tabetic' cases, in their symptoms bear considerable resemblance to locomotor ataxy. It is possible that the recovery of this form of cases under simple thermal treatment may have helped in giving some spas a reputation of being able to cure incipient cases of tabes dorsalis. Moreover, some of the sensory symptoms of tabes itself are probably often due to changes in the nerves or nerve roots, rather than in the spinal column, and these symptoms (notably the 'lightning pains') may occasionally be temporarily relieved by warm baths, as well as by other measures.

BOSTON MEDICAL
OCT 2 1917
LIBRARY

are more useful than the tepid ones. Other mineral waters are more frequently employed internally for gout, but in many delicate gouty persons no mineral waters need be drunk, the treatment being limited to tepid baths aided by climate and diet.

In many cases of chronic rheumatism, sciatica, and in stiff joints from gout or chronic rheumatism, douches, massage, and Swedish gymnastics form by far the most important part of the treatment, altogether superseding, in some cases, the simple hot baths.

In selecting a spa of this group much must depend on the ability in the local medical guidance and on the skill of the persons applying the douches, massage, and Swedish gymnastics. In other cases the accommodation, accessibility, situation, climate, and elevation above the sea-level must be considered in addition to the temperature of the waters. In the following pages the situation, altitude, temperature of the waters, etc., of most spas belonging to this group will be found.

Wildbad and Ragatz-Pfaefers have been placed first to serve as types, and the rest of the spas have been arranged in the political geographical order mentioned in the Preface, viz. :—Great Britain, Belgium, German and Austrian Empires, Switzerland, France, Italy, Spain, and Portugal. Bath, Buxton, Wildbad-Gastein, Schlangenbad, and Plombières might equally well have been given as types of this class of spas.

Wildbad (Würtemberg) lies at an altitude of about 1,323 feet in the deep valley of the Enz, a typical valley of the Black Forest, with lofty, rather steep, pine-clad slopes on both sides, up which long zigzag walks may be taken. The main direction of the valley is from north upwards to the south; the climate is bracing, and even in hot weather the nights are rather cool.

Wildbad, in spite of the great quantity of visitors who resort to it during the season, has not become too large, and has fairly well preserved its reputation as the type of 'Wildbäder,' or indifferent thermal baths. The temperature of the springs varies from 91·5° to 104·5° F. The Eberhards-Brunnen and the Königs-Brunnen are the most used for drinking, but there is naturally little difference between the different springs. Karlsbad or similar salts are added when a laxative effect is required, and when desirable the waters of other spas are drunk at Wildbad.

The chief reputation of Wildbad depends on its baths. There are two excellent bath-houses, the Great Bath-house and König Karls Bad, which both belong to the Würtemberg Government. The kind of bath chiefly used is the 'Wild-Bad,' an ordinary thermal bath in which the water bubbles up from a sandy floor, and is kept continually running off by the overflow pipe, so as to imitate a bath at a natural thermal fountain. There are likewise ordinary thermal baths, cold water baths (for which the cooled thermal water is used), hot air and vapour baths, electric baths, douches, and a set of Dr. Zander's medico-mechanical appliances for 'Swedish gymnastics.' Poor patients can have cheaper baths in the Katharinen-Stift. In the bath the surface of the body becomes covered with small bubbles, probably of nitrogen, but this phenomenon is not supposed to have any therapeutic significance.

The indications for Wildbad are those of thermal indifferent springs in general, namely, chronic rheumatic and gouty affections in feeble subjects, stiff joints from these affections or from the results of injury, convalescence from acute and chronic diseases, irritable functional nervous affections, nervous dyspepsia, some gynæcological affections and chronic skin eruptions. Cases of commencing chronic organic nervous affections, amongst which may be classed paralysis agitans, though its anatomical pathology is not yet known, likewise resort to Wildbad, as to other mild thermal waters, and are said sometimes to derive temporary benefit from their visit. In those exhausted from overwork or town life the bracing fresh mountain air and the necessary alteration in their mode of life doubtless play a chief part in the results obtained. For those who cannot or care not to walk uphill there are walks along the valley in both directions. The season lasts from May 1 to the end of September.

Access: Railway in about 24 hours, *viâ* Strassburg (or Cologne), Carlsruhe, and Pforzheim.

Accommodation: Good.

Doctors: Weizsaecker, Haussmann, De Ponte, Josenhans, and Teufel.

Ragatz-Pfaefers (Switzerland, Canton Saint-Gall). The baths of Ragatz and Pfaefers in Canton St. Gall are both supplied by the thermal waters of Pfaefers, the first medical account of which was written in 1535 by the famous Swiss physician Paracelsus, and dedicated by him to Johann Russinger, the liberal-minded abbot of Pfaefers.

Ragatz, at an altitude of about 1,700 feet, is a station on the railway from Sargans to Chur. It is situated on the south side of the valley on both banks of the Tamina, where this stream issues from a narrow defile to join the Rhine. The surroundings are very beautiful, and give scope for a variety of excursions. A funicular railway from Ragatz takes one up to the ruins of Wartenstein, about 1,000 feet above the town, overlooking the valley.

A walk of about three miles (in a south-westerly direction) up the romantic Tamina Gorge brings one to the thermal spring and bath-house of Pfaefers, which has an elevation of about 400 feet above Ragatz.

The waters of Pfaefers are indifferent thermal, and it may be mentioned that they, like the waters of Wildbad, are especially rich in nitrogen gas. Their temperature at the source is 98·6° F.; in the bath-house of Pfaefers, 93·5° F.; and in the wooden pipes by which the water is conducted to the baths of Ragatz, the temperature falls to 89°–93° F.

The patients who make use of the baths at Pfaefers lodge in the bath-house. The present building, which was commenced by the monks of Pfaefers in 1704, is naturally somewhat old-fashioned, and its position in the deep gorge is rather too confined and sunless. It is used chiefly by Swiss families, most persons preferring to live and take their baths at the modern spa of Ragatz. In Ragatz there are four excellently arranged bath-houses for the ordinary thermal baths, and a swimming bath supplied with thermal water. There are likewise arrangements for douches and electric baths, and complete apparatus for the cold-water treatment is shortly to be added. An institution fitted out with Dr. Zander's medico-mechanical appliances for 'Swedish gymnastics' has recently been opened.

As at other 'Wildbäder,' the waters in many cases are used for drinking as well as for baths. In former times patients used to remain for many hours at a time in the bath, and even have their meals brought to them there; but the average duration of a bath is now about half an hour; and in the same way, although very large doses of the water were formerly used internally, now only three to six glasses daily are recommended.

The indications for Ragatz are of course much the same as for other spas similar in climatic situation and in the nature of their waters. One may mention chronic rheumatism, the uric acid diathesis, many digestive and functional nervous disorders in the delicate and more irritable classes of patients. In chronic non-tuberculous articular affections, in sciatica and neuralgias, the judicious use of massage and Swedish gymnastics is often added. Like most thermal baths, these baths are employed in many chronic 'gynæcological' affections and in chronic cutaneous eruptions. In cases of slow convalescence from various diseases the climate, the music, and the cheerful spa life are of great assistance. Early chronic cases of various organic diseases of the nervous system are said often to derive at least temporary benefit from treatment at Ragatz.

The season at Ragatz lasts from the beginning of May to the end of October; that of Bad-Pfaefers from June to the middle of September.

Access: Viâ Bâle and Sargans, in about 25 hours.

Accommodation: Good.

Doctors: Jaeger, Bally, Dormann, Norström, and (in Bad-Pfaefers) Kündig.

Bath (England, Somersetshire). — The waters of Bath (altitude 100 feet), the 'Aquæ Solis,' or 'Aquæ Sulis,' of the Romans, are the only really hot natural waters of Great Britain. Their temperature is 104° to 120° F., and according to Attfield's analysis they contain 1·3 per mille sulphate of calcium, ·3 per mille of sulphate of sodium, ·2 per mille each of chloride of magnesium and common salt, and about ·1 per mille of carbonate of calcium and sulphate of potassium. The Bath waters are best classed in the indifferent thermal group. Very extensive remains of the Roman thermae exist. The city is beautifully situated, and owing to the surrounding hills the climate is mild and equable, so that the waters can be used all the year round. Spring and autumn are, however, the favourite seasons for a 'cure.'

Owing to a variety of causes, and partly merely to 'change of fashion,' Bath has lost much of the fame which it acquired in the eighteenth century: a fame in great part due to Beau Nash, and the fashionable guests whom the amusements, organised by him, attracted. Recently, however, all kinds of hydro-therapeutic appliances, and treatment by douches and massage, have been introduced, similar to those employed at foreign spas. Such methods are essential to the efficiency of a simple thermal spa, and will probably increase the number of visitors to Bath.

The 'Aix douche-massage' is given by two attendants working at once after the manner of Aix-les-Bains. The 'Nauheim treatment' for cardiac affections has recently been introduced at Bath, the Sprudel baths of Nauheim being imitated, and Dr. Schott's exercises being employed (see *Clinical Sketches*, October, 1895). There are inhalation rooms for affections of the pharynx and respiratory system.

The waters are used internally as well as externally, their internal action being, doubtless, similar to that of the indifferent thermal waters generally. It is not likely that the minute quantities of iron and arsenic contained in the water exert any therapeutical action, and still less likely that the nitrogen or recently discovered argon[1] and helium in the water exert any special effect, the anæsthetic effects caused by the inhalation of nitrogen

[1] Argon is chemically inert, and nearly all that is taken into the body can be found again in the expired air.

appearing to be due rather to the diminution of oxygen than to any action of the nitrogen itself.

The hot baths and the 'Berthollet' natural local vapour baths are useful in chronic gout, rheumatism, and some cases of sciatica and muscular pains. When the patient's joints are stiff, he can be lowered into a bath on a crane-chair. In psoriasis and chronic cutaneous affections the prolonged immersion often exercises a beneficial action on the skin.

In the results of lead poisoning, in functional nervous troubles, in painful menstruation, the thermal and hydrotherapeutic measures may be useful. It is hardly likely that any benefit which chlorotic girls obtain at Bath is due to the small amount of iron and arsenic in the waters. A military sanatorium for rheumatic and gouty complaints, or for the effects of wounds and accidents, might be erected here as at Teplitz in Bohemia, Barèges in France, &c. Bath is open for invalids all the year round.

The table water sold in bottles as 'Sulis Water' is the natural Bath water artificially aërated with carbonic acid gas.

Access: From Paddington Station in 2 to 3 hours.

Accommodation: Good, though it might be improved.

Doctors: G. A. Bannatyne, S. P. Budd, W. Carter, C. Coates, A. E. W. Fox, H. W. Freeman, T. B. Goss, F. K. Green, J. G. D. Kerr, and others.

Buxton (England, Derbyshire).—The waters of Buxton (temperature 82° F.) must be classed with those of Bath; but they are not so hot, and the climate of Buxton (altitude 1,000 feet) is more bracing than that of Bath. The situation of Buxton, and its position in the beautiful and interesting Derbyshire Peak District, attract a crowd of visitors as well as invalids to the spa. It was several times visited by Mary Queen of Scots.

The Buxton waters are still more weakly mineralised than the Bath waters, and contain, according to Dr. Thresh, only ·2 per mille of bicarbonate of calcium, and about ·1 per mille of bicarbonate of magnesium. Besides the simple thermal waters, there are very weak chalybeate waters (according to Lord Playfair containing about ·015 per mille carbonate of iron), and both are employed internally, the thermal waters in doses from four ounces to half a pint. The baths are given at the natural tepid temperature of the water for four to seven minutes; or, in the case of weaker persons, in whom the power of reaction is unsatisfactory, they may be given artificially heated to a temperature of 86°–100° F. The duration of the hot baths usually is from three to fifteen minutes. After the tepid baths, a walk to favour the reaction is advised, if possible. It is unlikely that the nitrogen in the waters of Buxton can exert any special therapeutic effect.

Buxton is supplied with douches, crane-chairs for lowering cripples into their bath, and with various hydro-therapeutic appliances. Temporary slight disagreeable effects are sometimes observed during a course of the waters here as at other spas. The treatment should be carried out under the guidance of a medical man.

Chronic gouty and rheumatic affections, and the stiffness in the joints resulting from them, are especially treated at Buxton. In the various conditions of weakness, produced by prolonged attacks of gout and rheumatism, Buxton is often eminently useful, partly, no doubt, through the influence of its bracing climate. Other conditions, yielding to simple thermal waters and hydro-therapeutic processes, may likewise be benefited here. The 'Nauheim treatment' of heart affections has recently been introduced at Buxton.

The season is from April to September. Buxton is open all the year round, but weather seldom permits invalids to take a course of the waters during the colder months.

Access: In 4¼ hours from London (St. Pancras Station).

Accommodation: Good.

Doctors: Robertson, Gifford Bennet, Arthur Shipton, Lorimer, Armstrong, etc.

Matlock Bath (England, Derbyshire).—Matlock Bath (temperature of the water 68° F.) is situated in a beautiful valley in Derbyshire on the left bank of the stream, but too much in a hollow for the climate to be bracing. The waters which, according to Dr. Dupré, contain about ·2 per mille carbonate of calcium and ·1 per mille sulphate of magnesium, are chiefly used for bathing. There are establisments here where hydrotherapeutic processes can be employed. Lumbago, sciatica, gouty and rheumatic joint affections, occasionally find relief.

Access: From St. Pancras Station (London) in about four hours.

Accommodation: Good.

Doctors: Holland, Moxon, Innes, Topham, &c.

Bakewell, likewise in the Peak district, has similar waters (temperature 60° F.) and an old stone plunge bath said to be Roman. The name Bakewell (Badequelle of Domesday Book) is derived from its spring. The neighbouring Stoney Middleton has also similar waters.

The tepid spring (about 73° F.) of Clifton (Gloucestershire), over which a building called the 'Hot-wells' used to stand, was formerly famous, but is little known at present.

Mallow (Ireland, County Cork), a station on the railway from Dublin to Cork, possesses the only thermal (sub-thermal) spring in Ireland. The waters must be classed in the indifferent group, being nearly pure. It was formerly much frequented by invalids.

Chaudfontaine (Belgium, Department of Liège), with a railway station 4½ miles from Liège, on the line to Aix-la-Chapelle, is beautifully situated, and possesses indifferent thermal waters (temperature 104° F.)

Schlangenbad (Germany, Prussian Province of Nassau).—This spa lies at an altitude of about 900 feet in a deep valley leading in a northerly direction from the Rhine, which is about five miles distant. Its situation, just at the bifurcation of the valley towards the north, renders the climate mild, though the air is sufficiently well ventilated. It is connected with Eltville on the Rhine by a steam-tram, and a good carriage road leads over the heights to Langenschwalbach.

The woodland and lower mountain scenery of the neighbourhood is unsurpassed, and miles of shady walks lead in every direction through the forest immediately surrounding the spa. A great variety of excursions may be made to spots in the Taunus Mountains and on the Rhine.

The waters are thermal indifferent, exceedingly soft, rich, like rain water or dew, in oxygen and nitrogen, and in the white glazed baths have a beautiful bluish tint. There are nine different springs, the temperature of which varies from 81·5° to 89° F. Everything that can be expected from simple thermal waters and rest in pure fresh air, amidst delightful scenery, can doubtless be obtained at Schlangenbad.

The guests include persons requiring rest after overwork or work in unhealthy surroundings, convalescents, and patients suffering from simple dyspeptic troubles, neurasthenia and functional nervous troubles, and women suffering from those chronic gynæcological troubles which are likely to be benefited by simple thermal waters. The waters have a special reputation in chronic skin eruptions, roughness of the skin, and slight degrees of ichthyosis. Massage is to be obtained in suitable cases. Whey and goat's milk whey are employed in some digestive troubles, and the chalybeate waters of the neighbouring Schwalbach are brought here for anæmic patients.

For the more robust types of cases spas possessing more active waters are usually more suitable than Schlangenbad. It is, however, sometimes used for an 'after cure' by patients coming from Karlsbad, Marienbad, Ems, Kissingen, etc. There is, perhaps, no other spa which exercises so soothing an influence on the nervous system.

Access: Railway *viâ* Cologne to Eltville; thence about five miles by steam-tram.

Accommodation: Very good. Rooms may be had in the bath-house.

Doctors: Baumann, etc.

Badenweiler (Grand-Duchy of Baden) is beautifully situated at an altitude of 1,425 feet in the lower part of the Black Forest, near the Swiss frontier. The town is sheltered in the north, east, and south by a semicircle of pine-clad mountains. West winds predominate, and the equable mild temperature renders the place a climatic health resort for delicate patients and those suffering from pulmonary affections. When more bracing air is required patients can reside at the more elevated 'Haus Baden' (1,750 feet). Walks on the surrounding slopes have been arranged for a 'Terrain-Cur.'

The temperature of the nine indifferent thermal springs of Badenweiler is 79° F.; they are used more for bathing than internally. Besides the beautiful 'marble bath' there is another large bath open to the air, and smaller baths for separate patients may likewise be obtained. Neighbouring remains of the ancient 'thermæ' show that the waters were used in Roman times.

The tepid baths of Badenweiler, aided by the climate, are used for much the same class of patients as are other 'Wildbäder.' Amongst the patients are convalescents, overworked delicate persons, chronic rheumatic and gouty cases, cases of chronic neuralgia, neurasthenics, and 'irritable nervous' persons. The baths are usually employed in cases where much excitation is undesirable, but sometimes it is convenient to artificially heat the water or to render the baths more excitant by the addition of common salt, a 'Mutterlauge' obtained from some 'Soolbad,' or otherwise medicate them. The season lasts from May 1 to October 1.

Badenweiler is much more used as a climatic health resort and for rest after active courses of mineral waters than for bathing purposes.

Access: In about twenty-one hours from London.

Accommodation: Very good in hotels and private villas.

Doctors: Neumann, Thomas, Kollmann, and Fränkel.

Liebenzell, about eight miles from Wildbad, is beautifully situated (altitude 1,100 feet) in a Würtemberg Black Forest Valley, and has thermal springs similar to those of Wildbad, but the temperature is lower (72° to 82° F.), so that the baths have often to be artificially heated. This spa enjoys a special reputation in the treatment of gynæcological affections. It is five miles from the railway station of Pforzheim.

The accommodation is more homely than at Wildbad, and the place is not much resorted to by English invalids or visitors.

Landeck (Prussian Silesia) possesses indifferent thermal springs, having a temperature of 66° to 84·2° F. The waters contain minute quantities of sulphide of sodium and sulphuretted hydrogen, and were, therefore, formerly classed in the sulphur group. The spa lies at an elevation of 1,470 feet in the mountainous country of Glatz, eighteen miles distant from the railway station of Glatz.

Warmbrunn (Prussian Silesia) is an old-established spa in the Hirschberg Valley, situated on the northern declivity of the Riesengebirge, 1,090 feet above the sea. It possesses indifferent thermal springs, having temperatures of 96° to 104° F., which were classed formerly as sulphur springs, because three of the five smell slightly of sulphuretted hydrogen. There is likewise a chalybeate spring, the 'Victoria-quelle,' with ·07 per mille bicarbonate of iron. The railway stations of Hirschberg and Reibnitz are about four miles distant.

Gastein (Wildbad-Gastein) in Austria (Duchy of Salzburg).—The place (altitude 3,310 feet) where the thermal springs arise is called Wildbad-Gastein (or Bad-Gastein) to distinguish it from Hof-Gastein (altitude 2,755 feet), which lies about five miles to the north, and is supplied through wooden pipes by the same waters.

There are eighteen thermal springs having temperatures ranging from 78·5° to 121° F.; and although peculiar electrical conditions are claimed for the waters, their action is probably merely that of indifferent thermal waters in general, aided by the mountainous climate.

The waters are chiefly used for baths, and enjoy an old reputation in the treatment of nervous affections of various kinds, as well conditions of merely functional origin as chronic affections due to organic changes in the nervous system, such as tabes dorsalis. In true cases of the latter class only a limited amount of benefit can of course be anticipated. The hotter baths are used for neuralgias. In gout, rheumatism, chronic metritis, and the remains of inflammation in the female pelvic organs, they are of service like other 'Wildbäder.'

The climate is sometimes of especial service in the treatment of convalescents and in patients coming for an 'after-cure' after treatment at Karlsbad, Marienbad, etc. Wildbad-Gastein lies in a very sheltered position.

Occasionally, as in some gouty cases and cases of nervous dyspepsia, the waters are used likewise internally.

The season lasts from May 1 to the end of September, but the months of July and August form the main season; and, though the accommodation is good, during these months the place is so crowded with visitors that it is impossible to get rooms unless ordered some weeks in advance. Excitable persons who cannot bear noise ought to stay at some distance from the waterfall,

Access: Viâ Zurich and Innsbruck, or Munich and Salzburg to Lend; thence by carriage to Gastein in about three hours.

Doctors: Gager, Schider, Wassing, Weingerl, and Wick.

Voeslau in Lower Austria (altitude 810 feet) is situated in a pretty country, on the railway about thirty miles south of Vienna. It possesses indifferent waters (temperature 75·2° F.) only used for bathing, and chiefly by ladies for functional nervous troubles, etc.

Teplitz (Teplitz-Schönau) in Bohemia.—This spa, which, since its recent union with the neighbouring village of Schönau, has been called Teplitz-Schönau, is the oldest spa in Bohemia. It lies in a broad open valley at an altitude of about 730 feet, and is sheltered on the north by the Erzgebirge, and on the south by the Mittelgebirge, of which the Königshöhe (870 feet) immediately overlooking the town is a projecting spur.

The town possesses a considerable commercial importance, which tends somewhat to modify its character as a spa.

The waters are weakly alkaline, containing but little carbonic acid gas, and may be classed amongst the indifferent thermal group (temperature 83° to 114° F.). In February, 1879, the supply at Teplitz was suddenly interfered with owing to the accidental tapping of a communicating spring in working a coalpit near Dux. It seemed at first as if the underground stream had been diverted from Teplitz, but on a new boring being made in the town the supply of water was re-established, and is now as plentiful as can be desired, though it has to be pumped up.

There are many different bath-houses in Teplitz, the most luxurious of which is the Kaiserbad belonging to the town; in nearly all of these bath-houses patients can likewise be lodged. Besides the ordinary thermal baths there are baths of peat ('moor-baths'), the peat being obtained for this purpose from the neighbourhood. The Teplitz peat contains much less iron, and is said to be less stimulating, than that of Franzensbad and Karlsbad, which two latter both derive their peat from the Franzensbad moor. The Teplitz 'moor-bath' is given at a higher temperature (about 99·5° F.) than those of Franzensbad and Karlsbad (about 89·5° to 95° F.), and exerts a greater anodyne effect. Massage may be obtained in suitable cases.

The patients who visit Teplitz are mostly sufferers from chronic rheumatic and gouty affections, sciatica and other neuralgias, or some functional nervous affections. Temporary improvement is said to follow the treatment in some cases of commencing tabes. The baths are likewise used in chronic cutaneous eruptions, and in wounds and ulcers slow to heal. There are Austrian, Saxon, and Prussian military hospitals at Teplitz.

The waters of Teplitz are, like those of other simple thermal spas, used more for baths than for drinking courses. The mineral waters of Karlsbad or Marienbad, to be obtained in the Kurgarten, may be employed in some cases, or those of the neighbouring alkaline springs of Bilin, or the bitter waters of Püllna, Sedlitz, and Saidschütz, near Teplitz. A stay at Teplitz is sometimes recommended as an 'after-cure' after treatment at Karlsbad, Marienbad, Franzensbad, etc., but Teplitz was probably formerly more used for this purpose than it has been recently, patients being now more often sent to the Alpine health resorts.

The season at Teplitz lasts from May to the end of September, but patients are received throughout the year.

Access: In two days, *viâ* Dresden.

Accommodation: Good.

Doctors: D. Kraus, Hirsch, Heller, Eichler, etc.

Johannisbad (Bohemia) lies at an altitude of about 2,300 feet in a mountainous region to the south of the Riesengebirge. Its waters belong to the indifferent thermal class, and have a temperature of 84° F. The effect produced in cases of prolonged convalescence, general weakness, and functional nervous disorders, is partly owing to the exhilarating nature of the climate. There is a chalybeate spring in the neighbourhood. Sometimes patients rest for a few weeks at Johannisbad after treatment at Karlsbad, Marienbad, etc. The season is from May 15 to the end of September.

Römerbad and **Tüffer** (Styria, Austrian Empire), both stations on the railway from Graz to Trieste, lie near to each other at an altitude of about 820 feet, and possess indifferent thermal waters (temp. 95° to 102° F.). Römerbad, like Schlangenbad, has a name for hysteria and chronic diseases of the uterus.

Tobelbad, an ancient spa in Styria (also called DOBBELBAD), lies at an altitude of 1,090 ft. Its two indifferent thermal springs have temperatures of 77° and 83·5° F. The railway station is twenty-five minutes distant.

Neuhaus in Styria, formerly called TÖPLITZ BEI NEUHAUS (altitude 1,200 feet), is a few miles from Tüffer and the railway station of Cilli. Its indifferent thermal waters have a temperature of 98° F. There is likewise a chalybeate spring.

Ofen or **Buda**, forming with Pest, on the opposite side of the Danube, the city of **Buda-Pest**, capital of Hungary, possesses indifferent (weakly mineralised) thermal springs, and likewise thermal sulphur waters, and commodious bathing arrangements. Ofen is, however, better known for the springs of cold 'bitter water' in its neighbourhood, some of which, such as the Hunyadi Janos and Franz-Joseph waters, are largely exported.

Amongst waters of the German and Austrian Empires which belong to the indifferent thermal group, the following may also be mentioned :

Wiesenbad and Wolkenstein (Warmbad near Wolkenstein), in the kingdom of Saxony, Villach, in Carinthia, with altitudes of between 1,400 and 1,600 feet, and Brennerbad (altitude 4,360 feet), at the top of the Brenner Pass in the Austrian Tyrol, possess waters having the relatively low temperatures of 72°–85° F.

Rajeczfürdö, formerly called Rajecz-Teplicz, situated in Upper Hungary, 1,374 feet above sea-level, one hour from the railway station of Sillein, possesses thermal waters (temperature, 91·5° F.) which contain minute quantities of iron and alum, but may be ranked in the indifferent thermal group.

Warmer than these waters are those of Krapina-Töplitz (altitude 530 feet, and temperature of waters 99·5° to 110° F.) and Topusko (temperature of waters 122° to 135° F.), both in Croatia, and of Daruvar in Slavonia (temperature of waters 104° to 117° F.).

Loèche-les-Bains (Louèche-les-Bains or Leukerbad) in Switzerland (Canton of Valais).—This spa has an altitude of about 4,600 feet, and is situated at the commencement of the Gemmi Pass, about three and a half hours' drive from Louèche-la-Souste, a railway station on the line from Lausanne to Viège.

The waters have been classed in the indifferent thermal group, though they have a mineralisation of 1·9 solid parts per mille (chiefly sulphate of calcium), and, like the waters of Bath and Bormio, may equally well be placed in the earthy group. The temperature of the springs is from 102° to 124° F., the Saint-Laurent spring being the warmest. There are about twenty different springs.

The climate is doubtless of great assistance to the balneo-therapeutic treatment, which consists chiefly in long and short baths. The waters are likewise employed internally in daily doses of one to five glasses, and have a diuretic effect, and a sedative action in some cases of gastric irritability.

The short baths are employed in the same class of cases as ordinary thermal baths, but the prolonged bath treatment forms a kind of speciality of this spa. Usually the spa guests are advised not to begin the baths immediately after their arrival, but to wait a few days; all fatigue of the journey has by that time passed off, and the patient has got accustomed to the place. The prolonged baths are taken at a temperature of 93° to 95° F. and last one to six hours. Ladies and gentlemen clothed in woollen garments bathe in large baths common to the two sexes, where they can take light refreshments and play chess, draughts, dominoes, etc., on floating tables.

July and August are the chief months for the cure, and at that time patients begin to arrive at the baths by five o'clock in the morning, and usually take a cup of tea, coffee, or chocolate in the bath. After the bath they go back to bed for half an hour or an hour, then take a short walk, and are ready at 11 o'clock for a proper meal. At about 3 P.M. the afternoon bath commences, and is likewise followed by a rest in bed. At six is the chief meal of the day, followed by music, etc., in the evening. The duration of the day's bathing is at first only an hour, and is gradually increased.

About the tenth or eleventh day the patients expect to see a skin eruption appear, which is called the 'poussée.' It is polymorphic, varying from a slight redness to a moist dermatitis, and may be accompanied by constitutional symptoms, loss of appetite, etc. According to de la Harpe, it lasts from ten to fourteen days, and is absent in 9 per cent. of the cases.

The prolonged baths are found useful in chronic skin affections, including eczema, psoriasis, prurigo (?), chronic urticaria, acne, etc. Their effect probably depends on the maceration of the surface epithelium, and the constant temperature of the water.

The season lasts from June 18 to September 30.

Access: By railway, *via* Dijon and Lausanne, to the station of Louèche-la-Souste; thence three and a half hours by carriage. *Accommodation:* Now good. *Doctors:* De la Harpe, etc.

Plombières, France (Department of Vosges).—The town (altitude 1,300 feet) is built on the banks of the Augronne in a valley of the Vosges Mountains. Extensive remains of Roman baths exist. The waters belong to the simple thermal class (77° to 155° F.), but contain minute quantities of arsenic. Some of the springs impart a peculiarly 'unctuous' sensation, due to the presence of silicate of aluminium. They are hence called 'Sources Savonneuses.'

The indications are those for simple thermal treatment in general. The waters are employed for drinking and inhalation, but chiefly for baths and douches. The inhalation rooms are on the Wassmuth system, first introduced from Germany into France at Mentone. Sometimes the chalybeate 'Source de Bourdeille' is employed internally instead of the ordinary thermal water. The Plombières season is May 25 to October 15.

Plombières has a good reputation in the treatment of gastralgia, nervous dyspepsia, chronic catarrhal enteritis, chronic diarrhœa, and functional nervous disorders, especially in arthritic subjects. It is largely to Napoleon III. that the spa owes its modern improvements.

Access: Viâ Laon, Reims, Nancy, and Epinal; in about nineteen hours. *Accommodation:* Good. *Doctors:* Bottentuit, etc.

Luxeuil-les-Bains in France (Department of Haute-Saône).—The small town of Luxeuil (altitude 1,300 feet) lies at the western foot of the Vosges mountains. Roman remains exist here as at Plombières.

The simple thermal springs differ in temperature from 108° to 125° F., and are used chiefly in the form of baths for the classes of affections usually benefited by simple thermal methods. The 'Source du Puits Romain' and the 'Source du Temple' (temperature 82° F.) are chalybeate springs. Season: May 15 to September 30.

Bains-les-Bains in France (Vosges) lies between Plombières and Contrexéville, and possesses eleven indifferent thermal springs (temperature 84° to 122° F.). It is a quieter spa than its neighbour Plombières. The waters have a reputation in hysterical complaints, and may be used in the class of cases in which simple thermal water treatment is likely to be beneficial.

Aix-les-Bains, in France (Savoie).—This spa has for convenience been classed in the sulphurous group. (See p. 213.)

Aix, in Provence (Department of Bouches-du-Rhône) is the 'Aquæ Sextiæ' of the Romans, and possesses two indifferent thermal springs, having temperatures of 91·5° and 98° F.

Néris, in France (Department of Allier).—Néris (altitude 1,150 feet), in the valley of the Cher, about three miles from the railway station of Chamblet, was known to the Romans, as the remains of its ancient *thermæ* testify. Its alkaline waters (·41 per mille of bicarbonate of sodium; ·38 per mille of sulphate of sodium) are so feebly mineralised that they are best classed in the simple thermal group.

Néris has an excellently arranged large establishment for baths, douches, massage, etc. The waters of the six springs have temperatures ranging from 102° to 127° F.; they are chiefly employed externally. Néris is one of the places where they sometimes use the prolonged baths. It has a great reputation in affections of the female pelvic organs, and in functional affections of the nervous system.

The season extends from May 1 to the end of September.

Access: By Paris and Montluçon; forty minutes' drive from the station of Chamblet.

Doctors: Caternault, Faure, de Grandmaison, Morice, Peyrot, de Ranse.

Mont-Dore (France) in the Auvergne. (See p. 201.)

Alet (altitude 650 feet) in France (Department of Aude), 22 miles from Carcassonne, possesses three simple thermal springs, the temperature of which is 68° to 86° F. It has likewise a cold chalybeate spring.

Campagne (France, Department of Aude), 4 miles from Alet, possesses feebly mineralised earthy springs, which may be classed in the simple thermal group (temperature 84° to 88° F.). The waters were formerly better known. Visitors have to lodge at Esperrazza, two miles distant.

Chaudes Aigues (France, Department of Cantal).—The village is situated at an altitude of 2,050 feet in a narrow valley about three hours' driving distance from the railway station of Saint-Flour. Its waters are feebly alkaline (·48 per mille of carbonate of sodium), and contain minute quantities of the iodide and bromide of sodium, but are best classed amongst the indifferent thermal waters. The temperature of the five principal springs ranges from 88° to 180° F. The season is from June 1 to September 15.

Dax (France, Department of Landes).—The town lies on the left bank of the Adour, and is a railway station on the line from Bordeaux to Bayonne, 32 miles from the latter town. Dax was the Roman 'Aquæ Augustæ Tarbellicæ,' and derived this name as well as its present name from its hot springs (Dax, De Aquis). These have temperatures varying from 88° to 140° F., and may be classed in the indifferent thermal group.

The mud of the 'mud-baths,' for which Dax has a great reputation, is obtained from the banks of mud left from the periodical inundations of the river Adour. These baths are used for chronic rheumatism, stiff joints, sciatica, uterine, and hysterical nervous affections. They are given at a temperature of 86° to 113° F., and occasionally even as high as 122° F.

The climate somewhat resembles that of Pau, but is slightly warmer. Dax is therefore occasionally used also as a climatic station, and is open all the year, forming a winter as well as a summer resort; the accommodation is satisfactory.

Access: Viâ Paris and Bordeaux; about 24 hours.

Doctors: Laranza, Dimulle, Mora, Raillard, etc.

Bagnères-de-Bigorre (France, Hautes-Pyrénées) has been described in the earthy group. (See p. 256.)

Ussat (France, Department of Ariège), lies at an altitude of 1,400 feet in the narrow valley of the river Ariège. Its weak earthy waters (temperature 88° to 106° F.) are chiefly employed for baths in cases of gynæcological and hysterical affections. They may be ranked in the indifferent thermal group. Ussat is a station on the railway from Toulouse to Ax, about 14 miles from the latter place.

Bagnoles-de-L'Orne (France, Department of Orne) lies at an altitude of 530 feet amidst the picturesque country called the 'Norman Switzerland.' Its weakly mineralised waters, having a faint odour of sulphuretted hydrogen, may be classed in the simple thermal group (temperature 81° to 84° F.). They are used in the same kind of cases as other tepid waters of this class. Bagnoles is a station on the branch line from Briouze of the railway between Paris and Granville.

Bormio, in North Italy, lies in the Upper Valteline valley (Stelvio route) near the Swiss and Tyrolese frontiers at an altitude of 4,620 feet (*i.e.* altitude of the old bath). The springs contain a small amount of bicarbonate of calcium, and of the sulphates of calcium and magnesium, but may conveniently be classed with the simple thermal waters (temperature 91° to 102° F.). They have a reputation in cases of chronic rheumatism, gout, and the uric acid diathesis, also in chronic cutaneous eruptions. Douches and mud baths are employed, as well as the thermal baths.

At Santa Catarina (see p. 194), about three miles distant, are chalybeate waters, which are sometimes used by patients residing at Bormio. The climate aids the cure in scrofulous and neurasthenic patients. Owing to the sudden fluctuations in temperature, warm clothing must be brought. The season lasts from June 1 to the end of September.

Access: From the railway station of Sondrio by diligence in 10 hours; from the station of Meran in 17½ hours; or from the station of Landeck in 22 hours. From Chur over the Albula or Julier pass to Samaden and over the Bernina pass to Bormio in 24 hours.

Accommodation: good.

Doctors.—There is generally an English doctor from the Riviera during the season.

Battaglia is situated in the Eastern district of the

Euganean Mountains of Upper Italy. The excavations in the rocks here are partly artificial. They are used, in a similar way to the better known caves of Monsummano, as vapour baths, with a temperature of 110° to 116° F. The four springs, having a temperature of 136° to 160° F., contain common salt and sulphate of calcium, but no sulphur, as they were formerly supposed to do; they are similar to those of Baden-Baden, but still more weakly mineralised, and are best classed amongst indifferent thermal waters. Chronic gout and rheumatism and arthritis deformans are treated. For bronchial catarrhs there is a special room for the pulverisation and inhalation of the waters. Local mud baths are likewise employed in a similar way to those of Abano and Acqui; and massage can be performed in suitable cases. The chief season is from the beginning of May to the middle of October, but the baths are open all the year.

Access: By railway *viâ* Macon, Turin, and Milan; or by Bâle and Milan; Battaglia is a station on the railway between Padua and Bologna.

Accommodation: Fair.

Doctors: Pezzolo, etc.

Monsummano (Italy, Province of Lucca) lies in the Val di Nievole, about half an hour distant from the railway stations of Pieve and Monte Catini. Here is a large cave, a natural vapour bath filled with steam arising from large surfaces of hot water. It is used in the treatment of chronic rheumatism, sciatica, neuroses, etc. The temperature in different parts of the cave ranges from 84° to 95° F. The cave was discovered in 1849, and the successful treatment of Garibaldi helped in giving it a reputation. There is accommodation for patients in Upper and Lower Monsummano and at the neighbouring spa of Monte Catini.

The season is from the middle of May to the middle of September.

Valdieri (North Italy, Piedmont) lies at an altitude of 2,700 feet in the valley of the Gesso, 5½ hours' distance south-west of the railway station of Cuneo. Of its thermal springs, the Sorgente San Lorenzo has a temperature of 156° F. The waters are used internally and externally; a slimy substance or mud, consisting partly of organised material, is collected from the bottom of the springs, and employed in the form of local or general applications to the skin, like the muds of Abano, Acqui, and Battaglia. The affections treated at Valdieri include skin diseases, chronic rheumatism, and scrofula.

Pré-Saint-Didier (Northern Italy, Duchy of Aosta), near Courmayeur, lies at an altitude of about 3,000 feet, and possesses weakly mineralised thermal waters (95° F.), used for bathing only.

Ischia.—This island, beautifully situated in the Bay of Naples, possesses indifferent thermal waters (temperature 100° to 163° F.), known from ancient times. The chief baths of Ischia are at Casa Micciola; natural vapour baths exist at Castiglione and elsewhere in the island. Sand baths and sea baths can be taken on the shore. The island and bathing arrangements suffered terribly from the earthquake of 1883.

Panticosa, in the Spanish Pyrenees (altitude 5,400 feet), is described amongst the sulphur baths on p. 245.

Fitero (Spain, Province of Navarra) possesses weakly mineralised waters (temperature about 117° F.), which have a reputation in the North of Spain for chronic rheumatism, etc. They contain apparently under ·5 per mille solid constituents, and may be classed in the indifferent thermal group.

Caldas-de-Oviedo (North of Spain, Province of

Oviedo) possesses indifferent thermal waters (temperature 109° F.), containing, like many waters of this class, a considerable quantity of free nitrogen gas.

Sacedon, or La Isabella (Spain, Province of Guadalajara), possesses thermal waters (temperature 84° F.), containing a total of about ·75 per mille solids, chiefly sulphate of calcium. These waters, which were known to the Romans and the Arabs, are here classed in the indifferent thermal group.

Caldas-de-Gerez (Portugal, Province of Minho) lies in the mountains of Gerez, and possesses very hot weakly mineralised waters, which contain carbonic acid gas and a little iron. The water runs into hollows cut into the rock, and is used in the form of hot baths for chronic rheumatism and neuralgias. It is likewise taken internally. The accommodation might be much improved.

CHAPTER VII

COMMON SALT, OR MURIATED WATERS

Salt baths have a more stimulating action than baths of plain water. The salt water soaks through the epidermis and acts as a chemical excitant to the nerve-endings in the skin; to this is partly due the special stimulating effect of sea water (the ordinary natural salt water for bathing) as compared with river water. In some sensitive skins too much irritation may be caused, giving rise to urticaria or increasing an eczematous eruption.

The 'Soolbäder,' or brine baths of Germany and other countries, to some extent take the place of sea baths in inland districts, and like sea baths are employed in scrofulous, rickety, and other cachectic conditions. The natural brine springs vary[1] in strength; and besides common salt contain, like sea-water, smaller quantities of many other salts. (See p. 13.) The stronger brines are often diluted with plain water for baths, and the weaker ones are artificially strengthened by the addition of a concentrated mineral water ('gradirte Soole'), or of a 'Mutterlauge,'[2] that is, the concentrated

[1] The degree of saturation with common salt is actually or almost reached by the brines of Droitwich in England, and Rheinfelden in Switzerland.

[2] The German term, 'Mutterlauge' (French, 'Eau-Mère') has been used in preference to the English 'Mother Lye' or 'Mother Water.' The English terms are hardly ever employed; indeed, on reading a

solution of salts—calcium chloride, etc.—left when most of the common salt has been made to crystallise out. (See under KREUZNACH, p. 88.) The different 'Mutterlauges' vary considerably in the relative proportion of their constituents, amongst which besides calcium chloride are the remnant of common salt and the chlorides of magnesium, potassium, strontium, and lithium, likewise bromides, iodides, etc.

Muriated springs sometimes contain excess of free carbonic acid gas, and to the presence of this gas in the gaseous warm salt baths ('Thermal-Soolbäder') of Nauheim and Oeynhausen, the mechanical stimulating effect of these baths is largely due. It is this effect which is made use of at Nauheim (*q.v.*) in the treatment of cardiac affections.

When taken internally salt waters (see p. 23) exercise a gently stimulating effect on the gastric [1] and intestinal mucous membranes, as well as rendering the contents of the bowel more fluid. They are therefore useful in constipation. The direct effect on the gastric mucous membrane is increased by the large amount of carbonic acid gas present in some of these waters.

By their action in aiding the digestion of albuminous materials they help to increase the general nutrition. Unless they are taken in quantities which produce catarrh

notice of Woodhall Spa, in England, we observed that the German word 'Mutterlauge' was used.

[1] The taking of a glass of warm muriated water may be euphemistically likened to the taking of a cup of salted bouillon or chicken-broth. A recent French writer would have us regard bouillons as 'solutions of ptomaines;' the same objection can at all events not be raised against simple salt water. Nor can any one say that the taking of the latter beverage necessitates the needless introduction of uric acid into the body, which Dr. Alexander Haig (*Brit. Med. Journ.*, 1894, vol. ii. p. 1299) gives as a disadvantage of taking soup.

of the stomach and intestines, they do not cause emaciation, and by this circumstance differ greatly from the sulphated waters. We have, on the contrary, often seen increase of weight in thin persons as well during as after well-arranged courses of muriated waters. Muriated waters are, therefore, *cæteris paribus*, in spare or emaciated persons preferable to alkaline sulphated waters. They are used in cases of anæmia where iron is badly borne, in cases of Indian cachexia, and in convalescence from infectious diseases.

The presence of iodides and bromides in some muriated waters, Wildegg in Switzerland, Adelheidsquelle at Heilbrunn, Woodhall Spa in England, Hall in Austria, Kreuznach, etc., has been supposed by some to exercise a special alterative action in various cachectic conditions, and even in syphilitic affections, though the quantities of iodide which can be taken in the form of mineral waters are of course absurdly small, when compared with the doses given in ordinary medicines for the latter affection.

Owing to their laxative effect and indirect favourable influence on the heart's action, they are useful in accelerating the abdominal circulation and removing engorgement of the liver and pelvic organs, and especially of the hæmorrhoidal vessels. They may thus be useful in dyspeptic conditions, in hæmorrhoids, and in some chronic uterine complaints; their effect on uterine fibroids is not generally admitted. Von Noorden (*Practitioner*, 1896) points out that in addition to their laxative effect these waters may be useful in quite opposite conditions, especially in 'mucous' affections of the lower bowels.

According to Von Noorden the waters of Homburg, Kissingen, etc., are useful in certain cases of hydrochloric hyperacidity of the gastric juice, especially in young men with gastric neurasthenia and gastric hyperæsthesia, and with a consequent dread of taking food freely; they may also be serviceable in occasional cases of gastric catarrh, chiefly in older persons, and more frequently in men than women.

The combined external and internal use of muriated waters is serviceable in tendencies to catarrh of the gastric, intestinal, and respiratory mucous membranes, and in tendency to rheumatic fever. (See NAUHEIM, p. 83.) In these cases the skin is 'strengthened' and becomes less sensible to slight changes in the temperature and moisture of the air. In bronchitis the waters may be inhaled (see p. 27), and make the secretion of the bronchial tubes less viscid, promoting expectoration. In emphysema and chronic bronchitis the good results of the treatment are probably partly due to their indirect effect on the heart's action and general circulation. The occasional occurrence of acute attacks of gout, when gouty patients commence a course at Wiesbaden, etc., can hardly be attributed to the sodium chloride contained in the water, since thermal baths of other groups may likewise cause temporary exacerbation of the malady (cf. pp. 35 and 299).

When used in chronic rheumatism and sciatica the hot springs are more beneficial.

The muriated sulphated waters of Brides-les-Bains, Leamington, Cheltenham, etc., are in their action somewhat akin to members of this group. (See Chapter XI.) In the present group, Droitwich, Nauheim, Kreuznach, Homburg, Wiesbaden, Kissingen, and Baden-Baden, have been placed first, as being amongst the best known and most representative spas; they have been somewhat more fully discussed than the other members of the group, which are arranged in the geographical political order adopted in the previous chapter.

Droitwich (England, Worcestershire).—Droitwich in England, like Rheinfelden in Switzerland, possesses most concentrated brine-waters. These waters contain, according to the analyses, 31 per cent. of common salt, that is about ten times as much as sea-water. It is impossible to sink in such water unless a weight be attached to the body; the specific gravity[1] is compared to that of the waters of the Dead Sea. The Droitwich water contains likewise about 5 per mille sulphate of sodium, and 1·3 per mille sulphate of calcium.

The country is very pleasant, but the town itself is not beautiful. Owing to the dissolving process which perpetually goes on in the underlying salt beds, buildings gradually sink, and the level of the ground is changing.

Owing to evaporation ordinary methods of heating the concentrated brine water lead to a precipitation of the salt. The water has therefore to be heated by the addition of hot water before being used for bathing purposes. The time of immersion in the warm baths is about twenty minutes; they are usually given at a temperature of 98° to 101°. They are employed in muscular rheumatism, sciatica, and in chronic rheumatic and gouty affections, and exert a tonic effect in convalescence

[1] Dr. A. Garrod found the specific gravity of a specimen of Droitwich brine to be 1·195. The specific gravity of Rheinfelden brine is 1·205 according to Bolley; that of the Dead Sea is said to range between 1·172 and 1·227. All these waters are practically saturated solutions, and so is also the 'Big Rapids' American Water, Michigan, U.S.A., advertised as being 'the strongest natural medicinal water known,' which, according to the *Lancet* (January 4, 1896, p. 40), has a total mineralisation of 33·8 per cent., and contains, in addition to common salt, some calcium chloride, magnesium chloride, and bromide of sodium.

from acute illnesses. The treatment may sometimes in gouty patients provoke an acute attack of gout, a fact which is occasionally observed at many other spas. According to Dr. R. Saundby (*Brit. Med. Journ.*, November 2, 1895) the 'Nauheim treatment' of heart affections is shortly going to be introduced at Droitwich in conjunction with the brine bath, modified to correspond with the baths of Nauheim.

The undiluted water, if taken internally, exercises a very disagreeable irritative and purgative effect.

The baths are open all the year round, but the summer months are preferred for treatment.

Access: From Paddington Station (London) in about five hours. *Accommodation:* Fair.

Doctors: Cuthbertson, Jones, Corbett, etc.

Nauheim, Bad-Nauheim (Germany, Grand Duchy of Hesse).—Nauheim lies at an altitude of about 400 feet to the east, and at the foot of a projecting spur of the Taunus range. The value of Nauheim as a spa was first made generally known in 1859 by the late Professor Beneke, but it is owing to the writings of the brothers Schott that the spa has of late years become so notorious in England.

The different springs at Nauheim vary much in their balneo-therapeutic qualities; four are used for drinking, and three for baths. The two used chiefly for drinking are the Kur-Brunnen and the Karls-Brunnen, which have lukewarm waters containing about 1 to 1½ per cent. common salt, 1 per mille chloride of calcium, and are effervescent with free carbonic acid gas. The Ludwigs-Brunnen is a weakly mineralised muriated alkaline gaseous water, useful as a table water, especially in dyspeptic troubles or for diluting the two first-mentioned waters. The fourth spring, the Schwalheimer-Brunnen (1·3 per mille common salt, ·7 bicarbonate of calcium, ·01 bicarbonate of iron), within easy reach of Nauheim, supplies a slightly chalybeate, gaseous water, which may be used as a table water, especially in anæmic cases. Both these waters are sold in bottles in all the hotels and pensions of the town.

The waters used for the baths contain about 2 to 3 per cent. chloride of sodium, 2 to 3 per mille chloride of calcium, some bicarbonate of iron, and much carbonic acid gas. The temperature of the waters is 82° to 95·5° F. Two of these springs rise in jets from the ground, and hence have been named respectively the Great and the Little Sprudel; they are rich in carbonic acid gas, one of them containing 1,340 cubic centimetres to the litre of water.

Different kinds of baths are given: a simple salt bath, the carbonic acid gas having been allowed to partially or nearly completely escape [these baths may be given at different temperatures, and strengthened, if necessary, by the addition of 'Mutterlauge;'] an effervescent bath (the Sprudel bath); and an effervescent wave or surf bath (the 'Sprudelstrom' bath). The latter (a speciality of Nauheim) is the most stimulating; the Sprudel water used for it is conducted direct from the spring into the bath.

There is now a separate bath-house for the simple salt baths. Besides the baths there are rooms for inhaling the waters and 'Gradirhäuser,' by which the patients can sit, as at Kreuznach, Kissingen, etc.

A great many different affections can be treated at Nauheim. Scrofulous and rachitic children, convalescents, patients with functional nervous disorders, those with chronic catarrhal affections of the respiratory and alimentary tracts are treated here as at other common salt water spas. In neuralgic affections the hot thermal baths are useful. The gynæcological affections likely to be benefited by salt baths can of course be treated at Nauheim. Bronchitic patients may inhale the waters or sit by the 'Gradirhäuser.'

In disorders of the digestive system, drinking the water of the Karls-brunnen plays a similar part to drinking that of the Elizabethen-Brunnen in Homburg. When the undiluted water of the Kur-Brunnen is likely to induce a catarrh of the bowels, it may be diluted, preferably with the Ludwigs-brunnen, according to Beneke's plan, and then is said to resemble the water of the Rakoczy spring in Kissingen.[1] The muriated waters are usually taken before breakfast, either diluted or undiluted, in amounts of from 5 to 30 ounces. The Schwalheimer-Brunnen and Ludwigs-Brunnen may be taken later in the day, and form agreeable table waters.

In lingering results of acute or subacute rheumatism the various baths are useful in promoting absorption of the remaining products of exudation in the joints, and, according at least to Beneke's views (1872), in promoting absorption of the lymph from the affected cardiac valves. By their general tonic action on the system they probably also help in counteracting any tendency to relapse.

The stimulating effect of the Nauheim 'Sprudel' baths on the circulation enables them to be given at a lower temperature than ordinary baths: this effect is due to a reflex action from the skin, which is stimulated by the combined action of the salts, the bubbles of carbonic acid gas, and, in the case of the 'Sprudelstrom' bath, by the movement of the water. The salt water soaks through the superficial layers of the epidermis, and acts as a chemical irritant to the nerve-endings in the skin, whilst the carbonic acid gas and the movement of the water act as mechanical stimulants.

[1] Patients, however, themselves have stated that they have found the action quite different.

It is indeed to the treatment of disorders of the heart and circulation, as systematically elaborated on Beneke's lines by the Brothers Schott, that Nauheim owes much of its present reputation. According to this method, the baths are employed in conjunction with gymnastic exercises, and the effect of prolonged courses of this treatment is said to resemble that of digitalis. By the Nauheim method of treatment, subcutaneous œdema and effusions into the peritoneum and pleuræ, which are associated with imperfect action of the heart, or even with commencing failure of compensation in valvular diseases, have often been successfully treated. The heart's action gains in strength and regularity, whilst the œdema and other signs of imperfect action gradually disappear.

Care is necessary in beginning the baths. Dr. Theodore Schott says it is advisable to begin with one per cent. salt baths free from carbonic acid, and at a temperature of 92° to 95° F., the baths lasting six to eight minutes, and being followed by rest. They should be omitted for a day at frequent intervals. The temperature at which the baths are taken may be reduced gradually, from day to day, until 85·5° F. in suitable cases is reached, whereas the proportion of the solids they contain and the time of immersion are slowly increased. Later on in the cure the 'Sprudel' bath may be commenced, and finally the still more stimulating 'Sprudelstrom' bath. The whole course should last six weeks or more.

The exercises devised by the Brothers Schott form a system of 'voluntary movements with resistance,' similar to P. H. Ling's Swedish system of 'Widerstands-gymnastik,' but differing from the exercises in which Dr. Zander's 'medico-mechanical' appliances (those used for voluntary movements) are necessary, by the fact that the 'Widerstand' or resistance is supplied not by the weight attached to a lever or pulley, but by the hand of the doctor or skilled attendant supervising the exercise. In dilated hearts the immediate result of about ten minutes' exercise is often a diminution in the superficial area of cardiac dulness. This diminution does not last, and it would be out of place here to discuss its therapeutic significance,[1] but what is much more important is the satisfactory result claimed to follow a prolonged course of this treatment in the class of 'cardiac' cases mentioned above. There appears to be considerable danger, however, of patients with heart disease and insufficient compensation of such a severe character that rest in bed is absolutely necessary, being injudiciously recommended to try the Nauheim treatment.

The doctrine of the mode of action of this treatment opens up very difficult questions. In many cases of deficiency in the heart's driving power (with or without mechanical defect by valvular disease) there are other factors as well, which help to bring about the general morbid condition in question. The kidneys and skin may not be acting properly, and there may be disorders of the digestive system, all of which may interfere with the general nutrition of the body, and of the heart, as well as of the other organs. The careful use of the baths and exercises may help in removing these troubles, and so indirectly as well as directly act favourably upon the circulatory system. A laborious investigation of the metabolism of the patients during treatment is needed for the further elucidation of the exact action of baths and exercises (both active and passive exercises) in cardiac affections. Such observations made on healthy individuals would not be sufficient, but they should be

[1] Dr. M. Heitler believes that spontaneous variations sometimes occur in the dulness of normal hearts. See 'Die Percussionsverhältnisse am normalen Herzen,' *Wiener klin. Woch.*, 1890, p. 787.

made on the patients themselves, whilst undergoing a course of the treatment, and on patients suffering from different forms of cardiac affection, both those accompanied and those unaccompanied by valvular lesions. Of great value for this purpose would be the regular daily examination of the urine, as to its total quantity, specific gravity, and richness in urea, uric acid, and albumen (if present), and lastly, also, as to the variations in its specific toxicity, during the treatment.

In certain cases the baths appear to act favourably, whilst the exercises do no good (see Dr. W. A. Sturge, *British Medical Journal*, 1895, vol. i., p. 527, and the paper by Dr. R. Saundby, *British Medical Journal*, 1895, vol. ii., p. 1081).

Dr. Schott thinks that in some cases, when digitalis has failed to act properly, it may act when combined with the Nauheim methods. At the end of a course of Nauheim treatment, and in order to further the beneficial result, he often recommends carefully graduated climbing exercise.

The Nauheim season lasts from May to the end of September.

Access: To Frankfurt in about nineteen hours; thence about one hour by train. *Accommodation:* Good.

Doctors: Schott, Abée, Groedel, Beste, Bode, etc.

Kreuznach (Germany, Rhenish Prussia).—Kreuznach (altitude 340 feet) lies on both banks of the Nahe, about ten miles from its entrance into the Rhine. The town proper is somewhat cramped and old-fashioned, and its drainage arrangements are said not to be quite satisfactory, but Bad-Kreuznach has roomy streets and villas. The latter constitutes the south-western portion of the town, lying partly on an island, partly on the right bank of the river, at the commencement of the narrower portion of the Nahe valley; it has a special railway station of its own, and patients may avoid the old town as much as they like. About one and a half mile southwards up the Nahe, in the narrower part of the valley, lies the village of Münster-am-Stein (altitude 380 feet), with similar mineral springs to those of Kreuznach. The bold porphyry cliffs of Rothenfels and Rheingrafenstein, with the ruins of Sickingen's Castle of Ebernburg and that of the Rheingrafen, make the scenery towards Münster-am-Stein very striking. The climate of Kreuznach is extremely mild; too hot for some persons in the height of summer. The hill-slopes in the neighbourhood are mostly vineyards, and do not afford the shady walks which might be desired for patients; one has to walk some distance on to the hills to reach woods which offer protection from the sun; shade amongst the trees may, however, be obtained in the Kurgarten, which is about to be enlarged in the direction towards Münster-am-Stein.

The waters of Kreuznach contain about one per cent. of common salt and about two per mille chloride of calcium, with minute quantities of the bromide and iodide of sodium, and in their therapeutic action resemble those of any other 'Soolbäder,' the iodide and bromide not being present in sufficient quantity to modify their action. The springs are numerous, but the cold Elizabethquelle is the spring chiefly used for drinking, and in the cold water the salt taste is not so disagreeable to the palate as it would be in the case of tepid springs. Two or three glasses are drunk, by preference on an empty stomach before breakfast, but naturally the dose varies according to the age and complaint of the patient. The baths are warmed to the required temperature, and usually strengthened by the addition of 'Mutterlauge,' that is, a strong solution of salts left when most of the common salt of the Kreuznach water has been made to crystallise out; the Kreuznach 'Mutterlauge' contains about 20 per cent. of chloride of calcium. Owing to the action of the 'Mutterlauge' on stone and porcelain, wooden tubs have to be employed for the baths. The supply of mineral water is so great that all the hotels and most of the houses are supplied with it.

In the Kurhaus are excellent newly built hot air and vapour baths, with arrangements for douches and massage; there is likewise an inhalation room, in which the air is charged by the Wassmuth method with the very finely pulverised mineral water, and where patients recommended to inhale the water may sit dressed in their ordinary clothes, protected by a loose outer oil cloth. Between Kreuznach and Münster-am-Stein are many 'Gradirhäuser,' where patients may sit on the side away from the wind; these 'Gradirhäuser' are high fences formed by bundles of twigs, through which the water is made to drip so as to concentrate it as a preliminary to heating it over a fire in the process of obtaining common salt and 'Mutterlauge;' as the water drips the impetus of the falling drops and any wind there happens to be, carry fine particles of the water into the surrounding air, which are inhaled by patients seated in the immediate neighbourhood, just as fine particles of salt water are inhaled by persons at the seaside.

Amongst the affections treated at Kreuznach the various forms of scrofula and rickets take the chief place. The Victoria Hospital is a charitable institution under the patronage of the Empress Victoria, where during the year about six hundred poor scrofulous and other children can stay for about four weeks, and besides the baths can receive operation or treatment if necessary. It resembles in effect seaside children's hospitals, such as those at Margate in England and Norderney in Germany.

Many patients come to Kreuznach for chronic catarrh or tendency to catarrh of the throat, nose, larynx, and bronchi. In this class of cases the inhalation room can be used, and the mild climate must help greatly in the results obtained, though higher altitudes are often preferable. Some chronic skin eruptions are benefited by the baths. On the subject of the possible utility of the calcium chloride of the 'Mutterlauge' in certain cases of recurrent urticaria, etc., see p. 335.

The spa is largely resorted to for chronic catarrhal and chronic inflammatory conditions of the female generative organs and the remnants of pelvic cellulitis. It is not seriously maintained in Germany at present that the waters have power to produce absorption of fibroid or other tumours of the uterus, but in gynæcological complaints, as well as in other complaints treated at Kreuznach, the physicians are ready to aid the action of the waters by ordinary well-recognised methods of treatment. The hot air and vapour baths are useful in the treatment of some obese patients. The season lasts from May 1 to the end of September.

Access: By train in about nineteen hours *viâ* Cologne and Bingerbruck, or *viâ* Metz. Patients should get out at the Bad-Kreuznach station.

Accommodation: Very good.

Doctors: Strahl, Hessel, Engelmann, Markwald, Prieger, Heusner, von Frantzius, etc.

Homburg, Homburg vor der Höhe (Germany, Prussian Province of Hesse-Nassau).—Homburg lies at an altitude of about six hundred feet, protected on the west by the Gross-Feldberg, Alt-König, and other heights of the Upper Taunus, and on the north partially sheltered by lesser heights. Owing to its open position the air is 'fresh.' The throng of visitors who resort to Homburg, as a place of amusement and fashionable society, cause its character as a health resort to be somewhat modified.

The springs are comparatively cold; those used for drinking are the Elizabethen-Brunnen, the Kaiser-Brunnen, the Ludwig-Brunnen, the Luisen-Brunnen, and the Stahl-Brunnen, all of which are rich in carbonic acid gas. The Luisen-Brunnen and the Stahl-Brunnen are muriated chalybeate springs containing much CO_2. The Stahl-Brunnen is the richest of the two in iron (about 5 per mille common salt, 1 per mille bicarbonate of calcium, and ·09 per mille bicarbonate of iron), and is compared to the Wein-Brunnen at Schwalbach; both the Stahl-Brunnen and the Luisen-Brunnen smell slightly of sulphuretted hydrogen, like the Pouhon of Peter the Great at Spa. The other three springs give forth sparkling muriated waters, *i.e.* the ordinary Homburg waters. These, in addition to common salt, contain small quantities of the chlorides of calcium and magnesium, and of bicarbonate of iron. The Elizabethen-Brunnen, the one most generally used, contains about 1 per cent. of common salt.

The baths now used are of metal, so arranged that the water can be warmed from a hot steam chamber at the bottom with the least possible escape of carbonic acid gas. There are inhalation rooms and douche arrangements.

Amongst the patients who resort to Homburg are those with gouty affections and the uric acid diathesis, some with habitual constipation, for whom the alkaline sulphated waters of Karlsbad and Marienbad are too strong, and patients with catarrhal affections of the alimentary and respiratory tracts. Chronic rheumatism is likewise often treated at Homburg. The same class of gynæcological affections are benefited at Homburg as at other salt water baths. The fashionable character of the spa renders it perhaps less suitable for scrofulous and rachitic children than other 'Soolbäder.' For overworked persons and those with functional nervous affections quieter spas are often to be preferred.

The iron springs of Homburg are employed for anæmia and debilitated patients, either alone or in conjunction with the ordinary muriated waters; they may be conveniently taken after meals, whilst the muriated waters are taken, when possible, on a fasting stomach before breakfast. Such matters, however, must be specially ordered by the spa doctor to suit individual cases.

There are well-known establishments for Swedish gymnastics, massage, electrical treatment, etc., at Homburg. The season lasts from May to the end of September.

Amongst many pleasant excursions which can be made from Homburg is the particularly interesting antiquarian one to the Roman fortress of the Saalburg, about four miles distant (see *Proc. Soc. Ant.* London, March 20, 1890).

Access: To Frankfurt in about nineteen hours; thence about half an hour by train.

Accommodation: Very good.

Doctors: Deetz, Weber, Hoeber, Will, Schetelig, Friedlieb, etc.

Wiesbaden (Germany, Prussian province of Hesse-Nassau).—Wiesbaden (altitude 380 feet), formerly the capital of the Duchy of Nassau, is a beautiful town with handsome public buildings and private villas, and well laid-out grounds, where patients and visitors can promenade. It is protected on the north by the Taunus range, and the climate is fairly mild. Though in the midst of summer the heat is very great, there are numberless shady walks to be enjoyed in the woods of the neighbouring Taunus Mountains. These woods are carefully kept up, less for profit than for the recreation of the inhabitants and visitors of the neighbourhood. A recently constructed funicular railway up the Neroberg (725 feet), whence a beautiful view is obtained over the surrounding country and distant hills, carries one at once right into the Taunus forest.

The waters of Wiesbaden were known to the Romans and were described by Pliny as the 'Fontes Mattiaci;' they are thermal common salt waters containing about 7 per mille of common salt. Their temperature varies from 100° to 156° F. The 'Kochbrunnen'[1] is the hottest spring, and the one probably most used for drinking. Other springs used for drinking are the Wilhelmsquelle, the Adlerquelle, and the Schutzenquelle. About twenty-four different springs are used for baths; the supply is abundant, and many of the hotels have their own spring and baths. A thin ochrous scum settles on the surface of the water when allowed to stand, and occasionally there is the very faintest smell of sulphuretted hydrogen.

Perhaps chief amongst the patients treated at Wiesbaden come those with chronic (atonic) gout and rheumatism. Chronic catarrh of the larynx and bronchi

[1] The water sold in bottles as 'Wiesbadener Gichtwasser' is a preparation made from the water of the Kochbrunnen, the main difference being an addition of about 8 per mille bicarbonate of sodium in the former.

is likewise often treated at Wiesbaden; inhalation chambers are provided for these cases. Some kinds of dyspepsia and chronic diarrhœa derive much benefit from drinking the waters. Chronic inflammatory conditions of the female generative organs are treated by baths as at Kreuznach. Syphilis is made a medical speciality of after the example of Aachen. At the new 'Augusta Victoria Bad,' besides the ordinary baths, there are elaborate arrangements for hot air and vapour baths, douches, compressed air baths (for pulmonary emphysema, etc.), massage, Swedish gymnastics and electrical treatment, which can be employed in suitable cases.

Frankfurt, Mainz, and the spas of Schwalbach and Schlangenbad can easily be reached; the amusements of visitors are well looked after; and it is no wonder that Wiesbaden is thronged with patients and visitors. The spa is open throughout the year, but the hottest weeks of summer are to be avoided by most people.

Access: By train in about 18 hours *viâ* Ostend and Cologne, or in about two hours longer *viâ* Calais and Cologne.

Accommodation: Very good.

Doctors: Conradi, Wibel, Ziemssen, E. Pfeiffer, Honigmann, and many others.

Kissingen (Bavaria).—Kissingen is beautifully situated in the fairly open valley of the Saale, at an elevation of about 600 feet above the sea-level. It is surrounded by wooded hills, where the so-called 'Terrain-Cur' can be taken, consisting in graduated gentle uphill and downhill exercise. The climate is mild.

Of the springs used for drinking the most important is the Rakoczy-quelle, which yields a cold effervescent water, containing about 6 per mille common salt and small quantities of the chlorides of potassium, lithium, and magnesium, and of the carbonates of iron (·03 per mille) and lime (1 per mille). The Pandur-Quelle is similar to the Rakoczy, but slightly weaker. The Max-Brunnen yields a pleasant, weakly mineralised, cold, effervescent water. All these three springs arise close together in the Kurgarten. The Rakoczy and Pandur waters, when a greater laxative effect is required, are recommended to be mixed with a product termed 'Kissingen bitter water,' obtained, according to a method of Liebig, from the 'Soole' springs. In catarrh of the stomach and bowels the water is warmed before drinking, though most of the carbonic acid gas escapes during the warming process.

The usual time for drinking the waters is the morning before breakfast between seven and nine, but a second dose is sometimes taken in the afternoon. Water from the neighbouring spa of Bocklet (see p. 186) is brought to Kissingen fresh every day, and may be given in cases when, owing to anæmia, a chalybeate water is considered suitable in addition to Kissingen water; it is usually taken later in the day than the Rakoczy and Pandur waters. The Max-Brunnen water is sometimes used merely as an agreeable aërated draught at various times of the day.

There are three bath-houses, two in the town and one in the valley close to the 'Gradirhäuser,' about 1½ mile to the north of the town; to the latter a small steamboat runs during the season. The springs chiefly used to supply the baths are the Salinen-Sprudel, close to the 'Gradirhäuser,' and the Schönborn-Sprudel, further off, at the village of Klosterhausen; the Pandur-Quelle is also used for baths.

Different kinds of baths can be supplied: firstly, ordinary Soolbäder at different temperatures; secondly,

Soolbäder rendered more stimulating by the addition of Kissingen 'Mutterlauge;' thirdly, 'Wellenbäder.' The water in the baths is heated by a coil of tubing containing steam, and this method avoids any unnecessary escape of the carbonic acid gas. The waves of the 'wellenbäder' are caused by a jet of water forced in through a hole in the bottom of the bath when the tap is turned on; with these baths are likewise provided a douche of soole water and a shower-bath of plain water. Mud-baths and ordinary douches may likewise be obtained.

Carbonic acid gas collected from the water is employed to supply carbonic acid gas baths, the application of the carbonic acid atmosphere being, of course, confined to the body, so that very little, if any, should be inhaled. There are likewise rooms for inhaling the waters in the form of a very fine spray; the inhalation is employed in the case of patients suffering from catarrhs of the respiratory organs, and such patients may likewise inhale the air near the 'Gradirhäuser.'

Patients are treated for various complaints at Kissingen; there are those suffering from hæmorrhoidal troubles and constipation, those with catarrhal conditions of the stomach or bowels, with or without a tendency to diarrhœa, gouty and rheumatic affections and functional nervous disorders, especially when supervening on an anæmic or scrofulous basis. The hot baths and hot mud-baths are useful in neuralgic pains. Massage can be employed in suitable cases. In some cases of anæmia, especially those with tendency to constipation, the waters of the Rakoczy spring appear to act more beneficially than iron in the form of medicines or in the stronger iron waters. Anæmic conditions with enlarged spleen after malaria are sometimes benefited.

Some chronic skin eruptions are benefited by the baths, and the same class of gynæcological affections are treated as at other salt and thermal spas; in these cases local treatment is, of course, often necessary, in addition to the balneotherapeutic treatment. Poor scrofulous children are treated at a charitable institution of the town. Bronchitic affections have been already referred to. In the treatment of glycosuria, obesity, the uric acid diathesis, and the slight forms of commencing nephritis, which occasionally come to Kissingen, the diet has, of course, to be carefully regulated, and in such cases it is sometimes an advantage if the patient can be treated, not in an hotel, but at an institution under the direct care of a resident physician.

The season lasts from May to the end of September. Patients are sometimes sent for an 'after-cure' to the neighbouring chalybeate spas of Bocklet or Brückenau, but, on the subject of 'after-cures,' the reader is referred to p. 43.

Access: In about 27 hours *viâ* Aschaffenburg and Würzburg.

Accommodation: Very good.

Doctors: O. Diruf senior, Stöhr, Sotier, Diruf junior, Dapper, and others.

Baden or **Baden-Baden** (Grand-Duchy of Baden).—Baden-Baden, so called to distinguish it from Baden in Switzerland and Baden near Vienna, was already known to the Romans as 'Civitas Aurelia Aquensis.' It lies at an altitude of about 650 feet, in a situation almost unrivalled for natural beauty, in the Oos Valley, close to the fertile plain of the Rhine. Though not completely sheltered from the north, its climate is mild with early spring and late summer.

There are over twenty different thermal springs; their

waters closely resemble each other in mineral constituents, and their temperatures vary from 124° to 150° F. The Hauptstollenquelle is the one most used for drinking, and contains 2 per mille common salt, ·05 per mille chloride of lithium, and a trace of arsenic, ·0007 per mille arseniate of calcium. The lithium has been said to exercise a special therapeutic action in gout, and the arsenic in skin affections; but it is probably more rational to regard the waters of Baden as simple muriated thermal waters, which, owing to their weak mineralisation, approach the indifferent thermal group of waters. If a laxative action be desired, Karlsbad, Marienbad, or Kissingen salts may be added to the Baden water.

The waters are much used for drinking, but still more for bathing. The Friedrichs-Bad and the Kaiserin-Augusta-Bad (the latter for ladies only) are amongst the finest, if not the finest, bath-houses in Europe. Here various kinds of baths may be had: ordinary thermal baths, strengthened by the addition of salt if necessary; the so-called 'Wild-bäder,' *i.e.*, baths with a sandy floor, as at Wildbad, in which the thermal water is kept continually running to imitate bathing at a natural thermal fountain; hot air and vapour baths (for the vapour baths the natural thermal water is used); all kinds of douches and electric baths. In the Friedrichsbad is a very complete set of Zander's medico-mechanical appliances for 'Swedish gymnastics.' There is an institution containing chambers for compressed air. 'Moor baths,' after the manner of Franzensbad, are shortly likewise going to be introduced at Baden.

The indications for Baden-Baden are: chronic gouty and rheumatic affections in delicate subjects; the results of injuries to bones and joints, chronic skin affections, etc., in which ordinary thermal baths are useful; catarrhal and nervous affections of the digestive organs in delicate people, in whom the more active waters of Karlsbad, etc., are contra-indicated. For convalescents, cachectic conditions from malaria, emphysema, and chronic catarrh of the respiratory organs, the climate is likely to prove favourable. For emphysema and chronic bronchitis the use of the compressed-air chambers is said to be useful.

The neighbouring walks are most suitable for a 'terrain-cur' in persons with weakly acting hearts from obesity, etc. On the hottest days cool walks in the dense pine forests can always be selected, and for those patients who are able to make excursions of some hours' duration the ruins of the old Castle of Baden, the porphyry cliffs close by, the ruins of Ebersteinburg, Schloss Eberstein, and Yburg are amongst the points on the neighbouring heights which may be visited.

Owing, however, to the beauty of the town itself, its handsome villas and hotels, the magnificent scenery around it, the numberless excursions which can be made, and the amusements it offers to the fashionable world, Baden must necessarily attract more ordinary visitors than patients. The chief season is from May 1 to the end of October, but there is also a winter season. Those invalids who do not bear heat ought to avoid Baden between the beginning of July and the middle of August. Baden is used as an intermediate station for those about to spend the winter in the south of Europe, and those returning from warmer regions to their colder homes.

Access: By railway, *viâ* Brussels or Paris and Strassburg, or *viâ* Cologne, in about 20 to 27 hours.

Accommodation: Good. Besides hotels and 'pensions,' there are two or three private sanatoria, where patients remain under the direct supervision of the physician.

Doctors: Schliep, Heiligenthal, Baumgärtner, Frey, von Hoffmann, Suchier, etc.

Woodhall Spa (England, Lincolnshire).—Woodhall Spa, noted for its muriated waters containing minute quantities of bromides and iodides, lies in a flat country at the borders of the fens. Owing to the flatness of the country, sea breezes reach it, and the Scotch firs in the neighbourhood contribute to the healthiness of the climate. In 1891 Professor Frankland found that the Woodhall water contained no free iodine, but minute quantities of iodides and bromides. It has, we believe, not been shown that these salts can be absorbed through the skin, though minute quantities can doubtless be absorbed when the waters are employed in the form of vaginal douches, etc. According to Professor Frankland's analysis of 1891 it appears that the Woodhall waters contain 19·5 per mille common salt, 1·27 per mille chloride of calcium, 1·14 per mille chloride of magnesium, 0·4 per mille bromide of sodium, ·02 per mille bromide of potassium, and only ·0075 per mille of iodide of potassium.

Amongst diseases in which benefit has been claimed by the use of these waters are rheumatic and gouty affections, catarrh of the respiratory and alimentary tracts, and 'biliousness;' also scrofula and rickets, cases of leucorrhœa in women, and some skin affections. Dr. Williams found the water useful in fibroid tumours of the uterus, which is in accordance with results which have been claimed for Kreuznach. A 'Mutterlauge,' made from the Woodhall water, can be used for strengthening the baths. The waters are likewise used for inhalation purposes in chronic nasal, pharyngeal, and laryngeal catarrh.

Access: From King's Cross station, in about four hours.

Accommodation: Good.

Doctors: C. J. Williams, R. Cuffe.

Ashby-de-la-Zouche (England, Leicestershire).—The mineral waters of this spa (altitude 400 feet) were discovered in 1805, during the working of coal-fields at a depth of 700 feet. Their natural temperature is 62° F., and according to Dr. B. H. Paul's analysis they contain 18·7 per mille common salt, 2·2 per mille chloride of calcium, 1·6 per mille chloride of magnesium, 2·5 per mille sulphate of calcium, and ·08 per mille carbonate of iron. They are used for giving brine baths (much weaker than those of Droitwich and Nantwich) at various temperatures.

The baths are employed in muscular rheumatism, in chronic rheumatic, gouty, and also scrofulous affections. Owing to the carbonate of iron in the waters, their internal employment might be useful in some conditions of debility. The salt would probably aggravate skin affections of an irritable nature.

Access: From St. Pancras station, London, in about three hours.

Accommodation: Good.

Doctors: Orchard, Stables, Williams, Barber, and Roe.

Malvern (England, Worcestershire) is supplied with brine from Droitwich. The Malvern Wells (especially St. Anne's Well) were formerly famed for their supposed special medicinal effects, which were doubtless due in part to the drinking of a nearly pure water in excellent air, and in part possibly to faith. Great Malvern lies on the eastern slopes of the Malvern Hills, at an elevation of about 520 feet above the sea. Malvern Wells and Little Malvern lie respectively two and three miles to the south of Great Malvern.

Nantwich (England, Cheshire).—Nantwich is situated in a pleasant well-wooded country. According to Frankland's analysis the waters contain about 21 per cent. common salt, 2·2 per mille chloride of magnesium, 1·9 per mille chloride of potassium, 6·5 per mille sulphate of calcium, and 5·0 per mille sulphate of sodium.

The brine baths were opened in 1883, and resemble those of Droitwich, but the baths are heated with steam instead of hot water, and are therefore given less diluted than at Droitwich. Cases of lumbago and muscular rheumatism often receive benefit from the treatment. The summer months are to be preferred.

Access: From Euston Station (London) in about four hours.

Accommodation: Satisfactory.

Doctors: Lapage, Munro, Mathews, etc.

The brine baths of STAFFORD are similar to those of Droitwich. At SALTBURN-BY-THE-SEA, in Yorkshire, baths are employed of brine conveyed from the brine wells at MIDDLESBOROUGH. At MIDDLEWICH, in Cheshire, there are brine baths. Other muriated waters in England are those of Filey, near Scarborough, Thorpe Arch (Boston) in Yorkshire, and Admaston, under the Wrekin in Shropshire.

Harrogate (England).—Some of the Harrogate waters might be mentioned in the muriated as well as the sulphurous group (see p. 217).

Llandrindod Wells and **Builth Wells** (Wales, Radnorshire).—These spas are described amongst the sulphur springs (see p. 219).

Llangammarch Wells (Wales, Brecknockshire).—This spa (altitude about 600 feet) possesses muriated waters containing chlorides of calcium, magnesium and barium. Dr. Dupré's analysis shows them to contain about 2·6 per mille common salt, 1·2 per mille chloride of calcium, ·3 per mille chloride of magnesium, and ·096 per mille chloride of barium.

Llangammarch is well situated, lying in a wide valley, but protected to the north and east by high hills. The air is most invigorating to those suffering from mental overwork. The waters are used externally and internally, and might be useful in some cases of dyspepsia, chronic gout, and rheumatism, especially where any emaciation is to be avoided.

Chloride of calcium has some reputation in chronic glandular affections, whilst chloride of barium, the speciality of the waters, is said to strengthen the heart's contraction whilst reducing its frequency. A system of treatment in heart affections after the model of the 'Schott treatment' at Nauheim has been adopted at Llangammarch.[1]

Access: In about seven hours from Euston Station.

Accommodation: Good.

Doctors: A medical man resides at the spa during the season. Dr. Bennett, of Builth, comes to give advice.

Bridge-of-Allan, Airthrie (Scotland, Stirlingshire).—This spa, which is three miles from Stirling, has a muriated spring containing about 5·4 per mille common salt, 4·4 per mille chloride of calcium, and ·5 per mille sulphate of calcium.

It stands in a sheltered position, and is a favourite resort of the people of Edinburgh. Three tumblers

[1] See 'Cases of Heart Disease treated by the "Schott" Method.' By Dr. Leslie C. Thorne Thorne. *Lancet*, January 4, 1896.

before breakfast is the usual dose of the waters. They are heated artificially before being drunk, and are said to be aperient in action. The waters have a reputation for dyspeptic troubles, some of which, Dr. Macpherson hints, are due to whisky and oatmeal.

Doctors: Compson, Fraser, Paterson.

INNERLEITHEN in Scotland, about five miles from Peebles, has similar waters. The muriated waters of BRIDGE-OF-EARN (PITKEATHLY), situated in a picturesque country, one and a half mile from Perth, contain free carbonic acid gas. 'Pitkeathly water' and 'Pitkeathly cum lithiâ' are waters prepared by the lessees of the Wells, and sold in bottles.

Mondorf, in the Grand Duchy of Luxemburg (altitude 650 feet), possesses a muriated water (temperature 77°) which, in addition to 8·7 per mille common salt, contains 3·1 per mille bromide of magnesium. The waters are used for drinking, bathing, and inhalation.

Oeynhausen, or **Rehme-Oeynhausen** (Germany, Westphalia).—The town lies in a broad fertile valley (altitude 230 feet) on the Werra, before it joins the Weser; the time by railway from Hanover is about one and three-quarter hour, and from Cologne four and a half hours. Oeynhausen, the newer portion of the town, is the name by which the spa is now best known. The climate is fresh and mild.

The chief remedial agents are three muriated springs, termed Bohrloch I., Bohrloch II., and Bohrloch III., very rich in carbonic acid gas, with temperatures of 77° to 91·5° F., containing, according to Professor Finkener, 31 to 34 per mille common salt. Bohrloch No. 1 (temperature 80·6° F.) contains about 32 per mille common salt, and 3 per mille each of the sulphates of sodium and calcium, and something like 1,000 volumes per mille of carbonic acid gas. The Bülow-brunnen are two cold muriated springs containing 34 to 80 per mille common salt. The bathing arrangements are very good, and by the mixing of the different waters together and heating them, if necessary, baths can be given at different temperatures and of various strengths in salt and gas.

So much do the remedial agents of Oeynhausen resemble those of Nauheim that it will not be necessary to repeat here what has been already said, when speaking of the latter spa, concerning indications for treatment. Owing to special treatment of heart affections, Nauheim has become much better known in England than Oeynhausen. The chief season is from May 15 to the end of September; but there is likewise a winter season.

Doctors: Lehmann sen. and jun., Oetker, etc.

Soden in the **Taunus** (Germany, Prussian province of Hesse-Nassau).

Soden lies at an altitude of about 450 feet, seven miles to the west of Frankfurt-am-Main, just at the foot of the Taunus mountains, which shelter it on the north. The climate is mild and equable. There are twenty-four different muriated springs, designated by numbers like those of Nauheim; they vary in amount of common salt (2·4 to 15 per mille), and in temperature (52° to 86° F.); some of them, as the 'Champagner-Brunnen' (6·5 per mille common salt), are very rich in carbonic acid gas; the 'Sool-Brunnen' contains 14·2 per mille common salt and comparatively little carbonic acid. These and the 'Warm-Brunnen' and 'Milch-Brunnen' (3 and 2 per mille common salt) are much used for drinking. Some of the springs contain an appreciable amount of iron; the 'Soolensprudel' (temperature 86° F.), used for 'thermal-soolbäder,' is rich in carbonic acid gas (1525 per mille volumes), and smells slightly of sulphuretted hydrogen. The bath-house is well fitted up for salt baths and gaseous salt baths, and there are inhalation rooms, on the Schnitzler and Wassmuth systems, for the use of patients with chronic laryngitis and bronchitis. Twenty minutes distant from Soden is the gaseous chalybeate spring of NEUENHAIN (·04 per mille bicarbonate of iron).

The spa is chiefly resorted to by Germans with chronic catarrhal affections of the air passages and emphysema. Other patients are scrofulous children and persons suffering from dyspeptic symptoms of catarrhal origin. Heart affections have not been made a speciality as at Nauheim. The season lasts from May to the end of September.

Soden is reached in half an hour by railway from Frankfurt. Accommodation is satisfactory.

Doctors: Otto Thilenius, W. Stoeltzing, M. Fresenius, H. Hughes, J. Koehler.

As somewhat similar to the waters of Nauheim, Oeynhausen, and Soden, may be mentioned the gaseous thermal muriated waters of HAMM, KOENIGSBORN, and WERNE, in Westphalia.

Salzschlirf (Prussian Province of Hesse-Nassau), a station on the railway between Fulda and Giessen, is situated at an altitude of 820 feet in a pleasant valley to the north of the Vogelsberg. The muriated spring, 'Bonifacius-Brunnen,' used for drinking and bathing, contains 10 per mille common salt, ·21 per mille chloride of lithium, and ·005 per mille iodide of magnesium, with a fair amount of carbonic acid gas. Some importance has been attached to the lithium and iodine of this water in the treatment of chronic gout, rheumatism, and the uric acid diathesis. The 'Tempel-Brunnen' contains less lithium, and the 'Schwefel-Brunnen' contains sulphuretted hydrogen gas; whilst the 'Kinder-Brunnen,' with only 4·3 per mille common salt, contains ·6 per mille bicarbonate of calcium, and ·76 per mille sulphate of calcium.

The gaseous muriated sulphated water of the neighbouring village of GROSSENLUEDER (containing 15·4 per mille common salt, 1·3 per mille sulphate of magnesium, and ·04 per mille bicarbonate of iron) is employed for its laxative action.

Kiedrich (Prussian Province of Hesse-Nassau), near Eltville, on the Rhine, at the foot of the Taunus Mountains, has a sanatorium, and the 'Kiedricher Sprudel'—a muriated spring—which, with 6·7 per mille common salt, contains ·5 of potassium chloride, ·75 of calcium chloride, and ·06 of lithium chloride. Some importance has been claimed for the lithium chloride in this water.

Schmalkalden (Prussian Province of Hesse-Nassau) lies at an altitude of 970 feet, on the south-western declivity of the Thuringian Forest, and is the terminus of a branch of the 'Werra' railway from Wernshausen. The earthy muriated water (temperature 63° F.) contains 9·2 per mille common salt, and about 3 per mille sulphate of calcium. The walks in the neighbouring woods might be used for a 'Terrain-Cur.'

Aachen or **Aix-la-Chapelle** in Rhenish Prussia.—This spa has for convenience been described under 'Sulphurous Waters' (see p. 211), but might equally well have been classified as having thermal muriated waters.

Münster-am-Stein (Rhenish Prussia) is situated at an altitude of 380 feet, about one and a half mile further up the Nahe valley than Kreuznach. The waters are similar, and it will be unnecessary to add anything to what has been already said under 'Kreuznach' (*q.v.*).

Accommodation is good, though arrangements are somewhat simpler than at Kreuznach.

Pyrmont (Germany, principality of Waldeck-Pyrmont) possesses muriated waters with 7 to 32 per mille common salt. (See under the Chalybeate Waters, p. 181.)

Arnstadt (Thuringia, principality of Schwarzburg-Sondershausen) is a summer resort in a sheltered locality at the northern border of the Thuringian Forest (altitude 920 feet). It possesses strong muriated (26½ per cent.) waters, used for brine baths (Soolbäder), and the 'Riedquelle,' a weak muriated spring (3·8 per mille common salt) used for drinking.

The affections treated here include rickets and scrofula, various pelvic complaints in women, etc. The season lasts from April to the end of September.

Frankenhausen (Thuringia, principality of Scharzburg-Rudolstadt) lies at an altitude of 370 feet, on the southern declivity of the Kyffhäuser, a little over nine miles from the railway station of Artern. It possesses a cold 25 per cent. brine used for baths, or, in a diluted form, for drinking. There is an establishment for the treatment of scrofulous children.

Koestritz in the principality of Reuss (altitude about 550 feet) is pleasantly situated in the Elsterthal, and is a station on the railway. The brine used for its 'Soolbäder' contains 22 per cent. common salt. Hot sand baths are likewise employed.

Salzhausen (Grand Duchy of Hesse) is situated at an altitude of 470 feet, at the southern foot of the Vogelsberg, one and a quarter mile from the railway station of Nidda. Of its weak muriated waters the strongest has about 1 per cent. of common salt, 5 per mille chloride of magnesium, ·07 per mille iodide of sodium, and a moderate amount of free carbonic acid gas. The water is used for drinking; for bathing a concentrated water is used, if necessary, strengthened by 'Mutterlauge' from Kreuznach or Nauheim. There are likewise a sulphur spring and a chalybeate one.

Salzuflen, or **Salzufflen** (principality of Lippe-Detmold), a station on the railway between Herford and Detmold, possesses a muriated water containing 4 to 9 per cent. common salt. There are 'Gradirhäuser' in the neighbourhood.

Salzungen (Germany, Duchy of Saxe-Meiningen) lies in the beautiful Werrathal, at the south-western declivity of the Thuringian Forest, about 780 feet above the sea-level. Its muriated waters vary in strength from 3 to 25 per cent. The baths are ordinarily made with a 3 to 6 per cent. muriated water, but they can be strengthened by the addition of Salzungen 'Mutterlauge,' which contains out of a total of 55 per cent. solids about 47 per cent. chloride of magnesium, 2·5 per mille bromide of magnesium, and 1·3 per mille iodide of magnesium. Douches, moor-baths, etc., can likewise be employed, and there are arrangements for inhalation treatment. The railway station is on the line between Eisenach and Meiningen. (Season, May 15 to end of September.)

Sulza (Thuringia, Grand Duchy of Saxe-Weimar) possesses muriated springs, having a total mineralisation of 5·3 to 14·5 per cent., and containing small quantities of iodine, bromine, and iron. The place, consisting of town, village, and salt works, lies between Weimar and Naumburg, on the Ilm, at an elevation of 480 feet.

Niederbronn (Germany, Alsace), the chief Alsatian bath (altitude 620 feet), is situated at the eastern declivity of the Vosges Mountains, on the railway from Hagenau to Saargemünd. Its two muriated springs contain about 3 per mille common salt, ·6 per mille chloride of magnesium, and ·01 per mille bicarbonate of iron (temperature 64° F.); they are chiefly used for drinking in cases of dyspepsia and catarrh of the bowels.

Rothenfelde is a 'Soolbad' in Hanover, situated at an elevation of 360 feet on the southern slope of the Osning-Gebirge. The muriated water contains a total of 67 per mille solids, 56 per mille being common salt, the rest consisting of chloride of magnesium, bicarbonate of calcium, etc. A concentrated brine, a 'Mutterlauge,' and a dried 'Mutterlauge,' and a weaker muriated water for drinking are likewise made use of. The 'Mutterlauge' contains 12·6 per mille bromide of magnesium. The railway station is on the line between Brackwede and Osnabrück.

Juliushall and **Harzburg** (Duchy of Brunswick, altitude 850 feet) lie close under the Burgberg. The cold muriated springs (there are two of them) contain between 6 and 7 per cent common salt.

These two places are much visited by the inhabitants of Northern Germany as summer health resorts, on account of their beautiful situations and the numerous excursions to be made from them in the Harz Mountains, to which they are adjacent.

Kolberg or **Colberg** (Pomerania), on the Baltic Sea, has, besides its sea baths, muriated waters containing 2·1 to 5·1 per cent. common salt, ·6 to 1·8 per mille chloride of magnesium, and 1·5 to 4·4 per mille chloride of calcium. There are sanatoria for the treatment of scrofulous children, etc.

Inowrazlaw (Prussia, province of Posen) is a railway station at one hour's distance from Bromberg, According to an analysis of 1875 its 'Bassinsoole' contains

30·6 per cent. common salt, and may therefore be compared in concentration to the brines of Droitwich and Rheinfelden.

Wittekind (Prussian province of Saxony, altitude 200 feet) is situated one and a quarter mile from Halle, on the Saale. It possesses a muriated water (3½ per cent.) which is mixed with aerated water for drinking, and can be strengthened with 'Mutterlauge,' or bath salt, for bathing. The bath salt obtained from the waters contains 239 per mille chloride of calcium, ·45 per mille iodide of aluminium, and 14·7 per mille bromides.

Elmen, or **Alten Salza**, in the Prussian Province of Saxony (altitude 150 feet), is situated near Gross Salza, on the railway forty minutes distant from Magdeburg. It is said to be the oldest 'Soolbad' in Germany, and possesses amongst other springs the 'Victoria-Quelle' (26 per mille common salt), used for drinking, and a spring used for bathing, containing about 5 per cent. common salt and ·6 per mille bromide of magnesium.

Koesen (Prussian Province of Saxony), a 'Soolbad' in the Saale Valley, at an altitude of 370 feet, near Naumburg, possesses a 5 per cent. brine (temperature 65·5° F.), and a sanatorium for scrofulous children from Berlin.

Neu-Ragoczi, or **Bad Ragoczi** (Prussian Province of Saxony), named after the famous spring in Kissingen, is one hour from the railway station of Halle, on the Saale, and possesses muriated springs containing small amounts (up to 1 per cent.) of common salt and much nitrogen gas. The waters are used for drinking and bathing, and the nitrogen is employed for inhalation purposes.

Thale (altitude 740 feet), in the Prussian Province

of Saxony, lies under the Rosstrappe, at the entrance to the beautiful Bodethal, in the Lower Harz Mountains. The neighbouring HUBERTUSBAD, on an island in the Bode, is supplied by the 'Hubertusbrunnen' with a muriated water containing 14·3 per mille common salt, and 10·7 per mille chloride of calcium. The railway station of Thale is terminus of a line from Quedlinburg.

Dürrheim (Grand Duchy of Baden), in the Black Forest, at an altitude of 2,300 feet, half an hour's drive from the railway station of Marbach, possesses a 26 per cent. brine, used for bathing. The place is a summer resort, and can boast of a hospital, the 'Amélie-Bad,' for scrofulous children, etc.

Canstatt (Cannstatt) and **Berg**, near Stuttgart, Würtemberg.—These towns adjoin each other, and are connected by a tramway with Stuttgart, of which town they practically form the north-eastern suburb. There are several springs, the most used of which are the Wilhelms-brunnen at the 'Kursaal,' and the Sprudel and Insel-quelle, situated between Berg and Canstatt on a small island of the Neckar. The springs yield earthy muriated tepid waters, fairly rich in carbonic acid gas (about 2 per mille common salt), which can be used for drinking and bathing in catarrhal affections of the digestive and respiratory organs.

The mild sheltered climate is of great assistance in some cases, but, unfortunately for its usefulness as a spa, Canstatt is more and more assuming the character of the manufacturing suburb of an important town.

For drinking purposes the laxative action of the waters may be increased, if desirable, by the addition of Karlsbad salt, etc., and for bathing purposes the waters may be heated to a suitable temperature. The season lasts from the beginning of May to the middle of October.

Hall, in **Würtemberg** (Schwäbisch Hall), is situated at an altitude of about 980 feet, and possesses a common salt water (' Soole ') which, by the addition of the concentrated Soole, or of the 'Mutterlauge,' may be used for 'Soolbäder' of different strengths. There is also a weak sulphur spring. The season lasts from May 1 to October 1.

Jaxtfeld, or **Jagstfeld**, in Würtemberg (altitude 450 feet), lies at the junction of the Jagst with the Neckar, 6 miles from the railway station of Heilbronn. It possesses a 26 per cent. brine used for bathing, and an establishment (Bethesda) for weak children.

Reichenhall, in the Bavarian Alps, close to the Austrian frontier, is situated in a well-sheltered valley at an elevation of about 1,560 feet above the sea. It is the terminus of a branch railway on the Munich and Salzburg line.

Of the twenty muriated springs the Edel-quelle (22 per cent. of common salt) and the Karl-Theodor-quelle are the most important, and their waters are mixed together for baths, which may be strengthened by the addition of Reichenhall 'Mutterlauge,' rich in chloride of magnesium. For internal use the water is diluted. A laxative 'bitter water' has been artificially prepared from the 'Mutterlauge.' For inhalation purposes there are the 'Gradirwerke' with the spray from a neighbouring mineral water fountain, and inhalation rooms where the finely pulverised water may be inhaled. There are arrangements for the inhalation of compressed air in emphysematous conditions.

Scrofula, rickets, and chronic catarrhal conditions of the respiratory organs are treated at Reichenhall. Walks exist for a 'terrain-cur' after Oertel's views in fatty infiltration of the heart, etc. The season lasts from the middle of May to the end of September.

Access: Viâ Munich and Freilassing, 4 hours from Munich.

Accommodation: Good.

Doctors: G. von Liebig, Rapp, A. Schmid, Cornet, etc.

Berchtesgaden, in Upper Bavaria, near the Tyrolese border, is a well-situated summer health resort, on the southern slope of the Untersberg, having an altitude of about 1,890 feet, and possessing a 26½ per cent. brine. Both the climate and brine baths can be of service in the treatment of rachitic and scrofulous conditions. Owing chiefly to its position, the place is used in the treatment of chronic affections of the respiratory organs, or for a rest after a course of waters at Karlsbad, Marienbad, etc. Berchtesgaden is very beautiful, but it is too much shut in by mountains to be called bracing. The season is from the middle of May to the middle of October.

Rosenheim (altitude 1,640 feet) in Upper Bavaria, at the junction of the Mangfall with the Inn, a station on the railway from Munich to Salzburg and Innsbruck, is a 'soolbad.' A mixture of the brines from Reichenhall and Berchtesgaden is used for the baths (the mixture constitutes about a 24 per cent. brine). There is also a weak chalybeate spring.

Dürkheim (altitude 380 feet), in Rhenish Bavaria, lies at the entrance to the Isenach-thal, on the western foot of the Hardt Mountains. Its muriated waters, which are mainly used for bathing, contain ¾ to 2 per cent. of common salt. The mineral waters of Dürkheim have the historical interest that in them the rarer metals, cæsium and rubidium, were first discovered, though, of course, only in traces.

Heilbrunn, in Bavaria, is a village situated at an altitude of 2,620 feet, 1½ hour distant from the railway station at Tölz. At Heilbrunn is the Adelheidsquelle, a cold muriated spring with 4·9 per mille common salt, ·9 per mille bicarbonate of sodium, ·05 per mille bromide of sodium, and ·03 per mille iodide of sodium. Amongst German muriated waters this is the richest in bromine and iodine; but it is doubtful whether these substances are present in sufficient amount to exercise any decided therapeutic effect.

Kreuth (Bavaria) possesses brine baths supplied from a distance; but it is doubtless the climate, due to its position in the Bavarian mountains 2,700 feet above the sea, which gives the place its chief value in the treatment of scrofula, convalescence, anæmia, &c. Kreuth lies in a sheltered position amidst beautiful woodland mountain scenery, between the Tegernsee and the Achensee, 2½ hours distant from the railway station of Gmund-am-Tegernsee. The Kreuz-quelle, a cold weakly mineralised earthy spring, containing a little sulphuretted hydrogen, is used for drinking and bathing. The season is June 1 to September 15.

Aibling, in the Bavarian Highlands, lies at an altitude of about 1,700 feet, and possesses a 'soolbad' supplied with brine from Rosenheim, and two weak earthy chalybeate springs. Moor baths, made up with the brine and 'Mutterlauge,' are employed in the treatment of scrofula, and the results of old inflammation in the pelvis or about the joints. Owing to its situation and elevation, the cooling of the air at dusk takes place very rapidly, and patients should be prepared for this. Aibling is a station on the railway from Munich to Salzburg.

Traunstein (Upper Bavaria), altitude 1,960 feet, a station on the railway between Rosenheim and Salzburg, has brine baths, for which brine from Reichenhall and Berchtesgaden is made use of.

Amongst German muriated waters, containing under 15 per mille common salt, the following have not yet been mentioned :—

SULZBRUNN, in Upper Bavaria, near the village of Sulzberg, at an altitude of 2,800 feet [some of the springs contain about 2 per mille common salt and ·015 iodide of magnesium]; SULZBAD, in Lower Alsace (3·2 per mille common salt; temperature, 60° F.); GANDERSHEIM, in the Duchy of Brunswick (waters contain up to 13·7 per mille common salt); SODENTHAL, or SODEN IN THE SPESSART VALLEY, in Bavaria (the waters have 9·4 to 13·8 per mille common salt and a small amount of bromide and iodide of magnesium); NEUHAUS IN BAVARIA, at the foot of the Salzburg, in the valley of the Saale (14·7 per mille common salt); and KÖNIGSDORFF-JASTRZEMB, in Prussian Silesia (about 11 per mille common salt, with a little bromide and iodide of magnesium).

Other German 'soolbäder,' with brines of various strengths (above 15 per mille common salt), are:— ADMIRALSGARTEN-BAD and other springs in BERLIN; OLDESLOE and SEGEBERG in Holstein; SALZDETFURTH in Hanover; CAMMIN and GREIFSWALD in Pomerania; ORB and SODEN-STOLZENBERG in the Prussian Province of Hesse-Nassau; SUDERODE, ARTERN, and DÜRRENBERG in Prussian Saxony; GOCZALKOWITZ in Prussian Silesia; WIMPFEN on the Neckar in Hesse-Darmstadt.

Ischl, in Austria (Salzkammergut), a station on the railway between Vienna and Salzburg, with mild and equable climate, lies at the junction of the Traun and the Ischl, about 1,500 feet above sea level. Some of the houses on the surrounding hills have a fresher air than the principal part of the town, which is built in a lower position on the banks of the Traun.

The Klebersberg-Quelle and the Maria-Luisen-Quelle, used for drinking, are weakly muriated (·5 per mille common salt), and for bathing there is a strong brine containing 23½ per cent. common salt, with a total of 24½ per cent. solids. The 'Schwefelquelle' contains, in addition to 17 per mille common salt, 4 per mille sulphate of sodium and a little sulphuretted hydrogen.

Moor baths, pine baths, and hydro-therapeutic treatment can be obtained at Ischl, which is likewise suitable for treatment by graduated exercise, the so-called 'Terrain-Cur.' The spa is much visited by Austrians. The season is June 1 to September 30. Owing to the beauty of the surroundings and good accommodation many persons visit Ischl more as a place of recreation and climatic health resort than on account of the waters.

Doctors: Max Mayer, etc.

Gmunden, in Upper Austria (Salzkammergut), lies at an altitude of 1,370 feet, on the beautiful Traun See, and is a station on the railway between Attnang and Ischl. The 24 per cent. brine from Ebensee is used at Gmunden for bathing. Season, June to October.

Hall in Upper Austria.—Bad Hall (altitude 1,060 feet) is a station on the branch line of the Kremsthal. It is celebrated for the muriated water of the Tassilo-quelle, which contains 12 per mille chloride of sodium, with ·058 per mille bromide of magnesium and ·042 per mille iodide of magnesium. This water, which is exported as the 'Haller Iodwasser,' was anciently known as the 'Haller Kropfwasser.' The exact therapeutic value of the iodine and bromide in the water remains doubtful.

There are likewise other salt springs. The Hall waters are used for scrofulous and rickety children, and for some gynæcological affections. There is a hospital for scrofulous children founded in 1855, a small sanatorium for poor adults, and a military sanatorium. The season is from May 15 to September 30. The bath salt extracted from the water contains 14·3 per mille chloride of calcium, 2·6 per mille iodide of magnesium, and 3·2 bromide of magnesium.

Doctors : Dr. Rabl, Dr. Koerbl, etc.

Hall in the Tyrol (Austria) not far from Innsbruck, is a climatic health resort, situated in the lower Inn valley, at an altitude of 1,700 feet. It is a station on the Munich and Innsbruck railway. The baths in the neighbourhood are supplied with a 24 per cent. brine, conducted from the Salzberg, about six miles distant. In the neighbouring village of Heiligenkreuz are a chalybeate spring and a weak sulphur spring. The season is from May 15 to September 30.

Aussee, summer health resort in Styria, at an altitude of 2,130 feet, is pleasantly situated in a broad valley of the Noric Alps, and possesses a 25 per cent. brine, used for bathing. Both the brine and the ‘Mutterlauge’ are employed in a diluted form internally. The railway station is 20 minutes distant. The season is May 15 to October 1, but the establishment of ‘Alpenheim’ is likewise open in the winter.

Hercules-Bad, near Mehadia, in Hungary. The ‘Hercules spring’ yields an unusually large quantity of thermal muriated water (temperature variable, 70° to 133° F.), free from sulphuretted hydrogen; but most of the springs are sulphurous (see p. 228).

Csiz, in Upper Hungary, in the Rima valley, three-quarters of a mile from the railway station. According to the analysis of Professor E. Ludwig, of Vienna, in 1890, the Hygiea-quelle, the only one used for drinking, is a muriated water containing a relatively considerable amount of iodine and bromine. In the *Bäder Almanach* it is advertised as the 'strongest iodine spring of the Continent.'

Also-Sebes (Hungary, in the Carpathian mountains, not far from Galicia) possesses muriated springs, which are used for drinking and bathing; the richest in common salt is the Ferdinands-Quelle (12·4 per mille). The Amalien-Quelle contains some bicarbonate of iron. The arrangements of the spa are unsatisfactory. It is 1½ mile from the railway station of Eperies.

Ivonicz (Galicia) in the Carpathians, at an altitude of 1,310 feet, possesses two gaseous muriated springs, the Karls-Quelle and the Amalien-Quelle (about 8 per mille common salt), which contain a little sodium carbonate (about 1·7 per mille) with iodide of sodium (about ·016 per mille); the first spring contains ·023 per mille bromide of sodium. There are likewise a chalybeate, a sulphur and a naphtha spring, the last of which is used for inhalation. Peat and mud baths are employed.

Salzburg, one of the most visited spas in Transylvania (altitude 1,590 feet), possesses brines of different strengths (from 5 to 15 per cent.), containing from 0·8 to ·25 per mille iodide of sodium.

Baassen (altitude 630 feet), picturesquely situated in Transylvania, 8 miles from the railway station of Mediasch, possesses muriated springs (temperature 54°–59° F.), containing minute quantities of iodide and bromide of sodium, and rich in carbonic acid gas. The Felsen-Quelle contains 4 per cent. common salt, ·03 per mille bromide of sodium, and ·013 per mille iodide of sodium.

Rheinfelden (Switzerland), an ancient town pleasantly situated (altitude, 866 feet) in the Rhine Valley in Canton Aargau, is protected by the Black Forest from north winds. It possesses a very strong 'Soole' or brine, containing, if the analyses be correct, about the same proportion of common salt as the Droitwich brine in England (*i.e.* 31 per cent.), or even a little more. The Rheinfelden brine (specific gravity, 1·205) is said to be the strongest on the continent, and is used in the form of baths for scrofulous and rickety children, and for patients suffering from the remains of old pleuritic or pelvic inflammation, etc.

Schweizerhalle (Switzerland) in the Canton of Bâle, situated on the left bank of the Rhine, at an elevation of 900 feet above the sea, is about 20 minutes' drive from the railway station of Pratteln. Its strong brine or 'Soole' contains, acording to Lunge, 30·7 per cent. of common salt; it is therefore practically saturated like the brines of Rheinfelden and Droitwich.

Bex (Switzerland, Canton Vaud) is a climatic health resort situated at an altitude of about 1,400 feet in the Rhone Valley, surrounded by mountains. It is a station on the railway from Lausanne to Brigue. The cold common salt waters (15 per cent. chloride of sodium) are used for ordinary 'Soolbäder' and various hydrotherapeutic processes. In a diluted form they may be used internally. There is likewise a cold sulphur spring sometimes used. The patients most treated here are scrofulous individuals with catarrhal tendencies. The 'grape cure' can be employed in suitable cases. Bex is open the whole year, but it is very hot in summer. The season is May to October. The daily fluctuations in temperature are greater than those at Montreux and other localities on the lake of Geneva.

Wildegg (Switzerland, Canton Aargau) is pleasantly situated in the valley of the Aar, about two and a half miles to the south of the spa of Schinznach (see p. 231). Its cold muriated water (10 per mille common salt, 1·6 per mille chloride of magnesium, 1·8 per mille sulphate of calcium) contains iodide of sodium (·028 per mille) and bromide of sodium (·013 per mille). This water is employed at the neighbouring spa of Schinznach in scrofulous affections, etc., and is exported.

Châtel-Guyon (France, Department of Puy-de-Dôme) lies in a pleasant part of the Auvergne, at an altitude of 1,300 feet. Its alkaline earthy muriated waters, rich in carbonic acid gas (temperature 82°–95° F.), contain a considerable quantity of iron, and, owing to the presence of chloride of magnesium, exert a laxative action. (Bicarbonate of calcium, 2·1 per mille; bicarbonate of sodium, ·95; bicarbonate of iron, ·06; bicarbonate of lithium, ·019; chloride of magnesium, 1·5; chloride of sodium, 1·6.)

Owing to the combination of the chlorides with the iron in the waters and their laxative character, the spa has been sometimes called the French Kissingen, and is more conveniently classed amongst the muriated than amongst the alkaline muriated group. The class of cases treated at Châtel-Guyon include dyspeptic and catarrhal conditions of the digestive organs and chronic constipation.

The distance from the railway station of Riom is about three and a half miles. The season lasts from May 15 to October 15.

Bourbon L'Archambault (France, Department of Allier).—This spa was once very famous, owing to the residence of Madame de Montespan and the visits of Catherine de' Medici, the Princes de Condé, Madame de

Sévigné, Madame de Maintenon, Racine, Boileau, etc., and, in later times, Prince Talleyrand. The town (altitude 870 feet) is a railway station on the line from Moulins to Cosne, about sixteen miles from Moulins.

The thermal muriated waters (temperature 126° F.), which contain 2·2 per mille chloride of sodium, are used in scrofula, chronic rheumatism, various nervous affections, &c. The 'Source Jonas' is a cold earthy chalybeate spring (·04 per mille of crenate and carbonate of iron) used for drinking only. The gaseous, cold, weakly mineralised water of the neighbouring spring of SAINT-PARDOUX likewise contains a little iron (·02 per mille of the crenate), and is used locally as a table water.

The season is May 15 to September 15.

Bourbon-Lancy (France, Department of Saône-et-Loire).—This spa, formerly famous through the visit of Queen Catherine de' Medici, possesses several thermal muriated springs (temperature 115°–133° F.) They are weakly mineralised (1·3 per mille of common salt), approaching the indifferent thermal group in character. The town (altitude 780 feet), which has a railway station, is thirty-one miles from Moulins.

Season: May 15 to September 15.

Bourbonne-les-Bains (France, Department of Haute-Marne).—Bourbonne is a small town lying in a pleasant valley at an elevation of 900 feet above the sea. It is a railway station on a branch of the line from Chaumont to Vesoul. The thermal muriated waters (temperature 109° to 150° F.) contain about 5 per mille common salt, about 1 per mille sulphate of calcium, minute quantities of other chlorides, and of bromide of sodium. They are used in the treatment of scrofula, chronic rheumatism, sciatica, &c., and chiefly in the form of hot baths and douches.

The 'Source Larivière,' a chalybeate spring in the neighbourhood, and the cold earthy spring of 'Maynard,' are sometimes likewise employed. The season lasts from April 15 to October 15. There are a civil and military hospital for balneo-therapeutic treatment.

Salins (France, Department of Jura).—Salins (altitude 1,200 feet) is a town situated on the stream Furieuse, surrounded by mountains. It possesses cold muriated waters containing 22 per mille common salt, ·03 per mille bromide of potassium, and traces of iodide of sodium. The waters are chiefly used for baths and douches; when used for drinking they are diluted and sweetened with a syrup. The baths are sometimes strengthened with the 'eau mère,' which contains 16 per cent. common salt, 6 per cent. chloride of magnesium, 8 per cent. sulphates of potassium, sodium, etc., and 2·8 per mille bromide of potassium. Salins is a spa for scrofulous and rickety children, convalescents, women with leucorrhœa, &c. It is the terminus of a branch line from Mouchard on the railway from Dijon to Pontarlier. The season is from June 1 to October 1.

Salins-Moutiers (France, Savoy).—This spa has been, for convenience, described with its neighbour Brides-les-Bains in Chapter XI. (See Brides-Salins, p. 163.)

Uriage (France, Department of Isère) is described in the sulphurous group (see p. 240).

Lamotte-les-Bains (France, Department of Isère) lies at an altitude of about 2,000 feet, and is a station on the railway from Grenoble to La Mure. It possesses thermal muriated waters (temperature 135° to 140° F.), which contain about 3·5 per mille common salt, some earthy salts, etc.

Salies-de-Béarn (France, Department of Basses-

Pyrénées) is a railway station on a branch of the railway from Toulouse to Bayonne. Its strong muriated waters contain about the same proportion of common salt as the brine of Nantwich in England (that is, about 21 per cent.), and ·16 per mille bromide of sodium. The 'eau mère' left after the extraction of the greater part of the common salt contains about 21 per cent. common salt, 15 per cent. chloride of magnesium, 5 per cent. chloride of potassium, and 1 per cent. bromide of magnesium. The brine baths are warmed before use.

Biarritz, the fashionable marine spa and winter resort on the south-western coast of France, now possesses baths supplied by the muriated springs of Briscous. The 'eaux mères' can be employed like the 'Mutterlauge' of Kreuznach, etc.

Balaruc (France, Hérault), standing at the edge of the salt lake of Thau, possesses three common salt springs, whose temperatures are respectively 54°, 66°, and 118° F. The principal spring, 'Source des Romains,' is the hottest, and contains about 7 per mille chloride of sodium. The waters are used for drinking, and for baths and douches in scrofulous and rheumatic affections. Salt mud baths are also used. Balaruc has an old reputation in torpid nervous affections; but comparatively few organic nervous diseases are now sent thither for spa treatment.

Abano, North Italy (Province of Venice).—Abano (about 100 feet), a few miles by rail from Padua, lies in the region of the Euganæan hills. The water of the thermal springs (temperature 100° to 183° F.), according to R. Nasini's analysis of 1894, contains 3·4 per mille common salt and about 1 per mille sulphate of calcium, and so little sulphuretted hydrogen that they can hardly be classified amongst the sulphur waters.

Protococcus and other primitive forms of plant life grow in the water. The springs were well known as the Aquæ Aponenses or Aquæ Patavinæ in Roman times, and in the sixth century Theodoric the Great ordered the baths to be restored.

Besides thermal mineral water baths, baths of a mud, rich in organised material and impregnated with the salts of the mineral waters, are employed, as at Acqui and Battaglia, in the form of local applications especially. Amongst the affections treated are rheumatic and gouty troubles and neuroses. The waters can be obtained also in Venice. The season is June 1 to September 30. Abano possesses a mild winter climate.

Monte Catini, in Italy (Province of Lucca).—The spa (altitude 920 feet) is prettily situated in the Val di Nievole. It possesses several thermal muriated springs (temperature 70° to 88° F.), containing 4 to 18 per mille common salt. The waters are employed internally and externally for dyspeptic conditions, scrofula, chronic rheumatism, etc. The season is from May 1 to September 30.

Castro Caro, in Italy (Province of Toscana), one hour's drive from the railway station of Forli, possesses muriated waters containing iodides and bromides. According to Professor L. Guerri, the water of the 'Sorgente Magnani' contains 44 per mille common salt, ·197 per mille iodide of magnesium, and ·185 per mille bromide of magnesium. The accommodation is comfortable.

Castellamare, in Italy (the Roman Fontes Stabiæ), through its beautful situation on the south coast of the Bay of Naples, is one of the most delightful marine spas of Italy, but it possesses also earthy muriated waters, employed from ancient times. The 'Acqua del Muraglione' is said to contain about 5 per mille common salt and 1 per mille bicarbonate of calcium. One of the springs contains iron, and another smells of sulphuretted hydrogen.

Caldas-de-Montbuy, in Spain (Province of Barcelona), possesses thermal muriated springs (temperature 153° to 158° F.) of great local repute for rheumatism, sciatica, and old wounds. The spa is a two hours' drive from the railway station of Mollet, and is resorted to before and after the hottest season of the year.

Cestona-Guesalaga, in the North of Spain (Province of Guipuscoa), owns thermal weak muriated waters, containing a little sulphate of calcium and sulphate of sodium.

Caldas-de-Malavella, in Spain (Province of Girone), possesses thermal muriated waters (temperature 145° F.), containing a total of only about 1 per mille solids (chiefly chlorides of calcium and magnesium). In fact, they might be likewise classed in the simple thermal group.

Ciechocinek, in Russian Poland, two miles from the Prussian border, possesses muriated waters (18 to 44 per mille common salt) and 'Gradirhäuser,' which can be used for inhalation.

CHAPTER VIII

SIMPLE ALKALINE WATERS

SIMPLE alkaline waters (see p. 23) are often of use in cases of dyspepsia in the more robust types of persons, especially when their condition is associated with catarrh of the stomach and intestines. Prolonged courses of simple alkaline waters are, however, injurious in catarrhal conditions of the digestive organs in weak people when the gastric secretion is deficient in acidity; such 'atonic dyspepsia' often occurs in anæmic and chlorotic persons, in convalescents from infectious diseases and persons debilitated from various causes. The muriated alkaline waters are to be preferred when any emaciation is to be avoided.

Simple alkaline waters exert a diuretic influence, and the alkali aids the action of simple water in increasing the other secretions of the body. They are useful in some cases of gout, and gouty glycosuria, and the uric acid diathesis; in tendency to gallstones and 'abdominal plethora.' Their action in cases of obesity has been ascribed to their combining with fat in the body to form soluble soaps, which are then excreted. This theory has not been confirmed, and the action of alkaline waters in obesity and glycosuria is probably largely, if not wholly, due to the accompanying regulation of the diet.

Alkaline baths have a more cleansing action on the skin than plain water baths of the same temperature; when much carbonic acid gas is present, they may exercise a mechanically stimulating action, similar to that of baths of other gaseous waters. Warm alkaline baths, like most warm baths, are useful in leucorrhœa and catarrhal conditions of the female pelvic organs.

Amongst the simple alkaline waters those of Vichy and Vals will be placed first as well-known types. The rest will be given afterwards in the political geographical order previously made use of.

Vichy (France, Department Allier).—Vichy, one of the most frequented spas of France, is situated (altitude 736 feet) on the right bank of the River Allier amidst a rather flat cultivated country of somewhat uniform aspect. Its fame as a fashionable spa dates from about the time when Madame de Sévigné (1678) came to undergo the treatment, and wrote such amusing letters about it.

The thermal alkaline waters of Vichy may be regarded as the representatives of the simple alkaline class. The different mineral springs of Vichy differ from each other chiefly in their temperature, in the proportion of free carbonic acid gas, and in the fact that some of them contain a mentionable amount of iron. Arseniate of sodium is found in traces (·002 per mille in the ‘Source Grande Grille’).

The ‘Grande Grille,’ the best known of the Vichy springs, has a temperature of 108·5° F., and contains 4·8 per mille bicarbonate of sodium. The ‘Source de l'Hôpital’ has a temperature of only 89° F., and contains rather more carbonic acid gas and bicarbonate of sodium (5 per mille). The three ‘Sources des Célestins’ are all cold; the ‘Source ancienne’ contains about the same amount of carbonic acid and bicarbonate of sodium as the 'Hôpital' spring just mentioned, whereas the 'Source nouvelle' and the 'Source de la grotte' are characterised by containing respectively ·044 and ·028 per mille of bicarbonate of iron. The spring of 'Mesdames,' which is conducted from its source (near Cusset about two miles distant) to the Établissement des Bains at Vichy, resembles the two foregoing springs, and contains ·026 per mille bicarbonate of iron. Amongst the other springs at Vichy the 'Puits Chomel' should certainly be mentioned; it is very similar in temperature and constituents to the 'Grande Grille,' but is less rich in carbonic acid, and slightly richer in bicarbonate of sodium.

The patients mainly treated at Vichy are dyspeptics, and persons suffering from gall-stones, the uric acid diathesis, and various gouty, hepatic, urinary, and female pelvic disorders. Sufferers from chronic rheumatism likewise often resort to this spa. All patients should have a certain amount of reserve strength, those with considerable cachexia of all kinds being unsuited to Vichy treatment, a point especially to be considered in the case of gouty and diabetic patients.

The waters of Vichy taken in suitable doses have, as a rule, no laxative action like the sulphated alkaline waters of Marienbad, etc., but they are nevertheless used in the treatment of obesity; in such cases of course the diet is most important, but it has been supposed that the alkaline constituents of the Vichy water form soluble soaps with the fats of the body, and thus favour their removal by excretion; this theory concerning the alkaline treatment of obesity is, however, doubtful.

In anæmic conditions the 'Source Mesdames' is the one generally preferred. The hot 'Grande Grille' has a special reputation in biliary calculi and hepatic complaints, the more strongly alkaline 'Source del'hôpital' being often preferred for gastric complaints, and the colder and more diuretic water of the 'Source ancienne des Célestins' for urinary affections. In recent times, however, the view of this sharp therapeutic division in the uses of the different springs has been considered somewhat arbitrary, and the doctors do not always act according to it. The 'Puits Chomel' was until lately mainly reserved for gargling, etc. in cases of chronic pharyngitis and respiratory affections, but its use has been lately extended to cases formerly treated by one of the other hot springs of Vichy.

The Vichy waters are taken at first in small doses, such as half a glass twice daily, and afterwards the dose is gradually increased to four or five glasses in the day; in former years much larger doses were given. The time preferred for taking the waters is about two hours after meals or an hour before; the more chalybeate Vichy waters ('Source Mesdames,' etc.) are often ordered to be drunk at meal times.

Baths are very much used at Vichy, in addition to the internal use of the waters. The large 'Etablissement des Bains' provides thermal mineral water baths, vapour baths, douches, and inhalation rooms. Besides the large establishment (first and second classes) there is a smaller one near the 'Source de l'hôpital.' Quite recently, in addition to other hydro-therapeutic treatment, massage with douching after the method of Aix-les-Bains has been introduced at Vichy.

The mineral water of Vichy is likewise employed for perinæal, vaginal, and rectal douches. In specially severe gastric cases the stomach is periodically washed out by the ordinary syphon method.

There are several alkaline mineral springs in the neighbourhood of Vichy, amongst which the various springs of Saint-Yorre, about five miles to the south of Vichy, may be specially mentioned; they are strong, cold alkaline springs, belong to private owners, and their waters are used for exportation.

The alkaline waters of Vichy and its neighbourhood are largely exported, and can be well used at home. The colder springs are better adapted for this purpose than the hot springs; those of Saint-Yorre, Hauterive, and Célestins may be specially mentioned.

The Vichy season lasts from May 15 to the end of September, but the baths remain open all the year. For most invalids the middle of summer is unpleasantly hot at Vichy. On the subject of 'after-cures' the reader is referred to p 43.

Access: In about 18 hours, *viâ* Paris.

Accommodation: Very good.

Doctors: Durand-Fardel, senior and junior, Cornillon, Willemin, Déléage, Glénard, Cormack, Jardet, etc.

Vals, in France (Department of Ardèche).—Vals is picturesquely situated at an altitude of 300 feet in a narrow valley watered by the Volane. It has been called the 'Cold Vichy,' and is celebrated for its remarkably strong cold alkaline springs, which are rich in carbonic acid gas, and contain 3 to 7 per mille bicarbonate of sodium and small quantities of the bicarbonates of lithium and iron. The 'Source Souveraine' contains ·042 per mille bicarbonate of lithium. Some of the springs (as the 'Rigolette') contain a considerable quantity of bicarbonate of iron, and the 'Vivaraise' spring No. 1, which has ·05 per mille of the bicarbonates of iron and manganesium, contains only 2 per mille of the bicarbonate of sodium. The number of the alkaline springs at Vals is very great, and as at Saint-Yorre, near Vichy, a cold alkaline water can be obtained everywhere by sinking a well to a certain depth. In the treatment of dyspepsia and cases of gastric catarrh, etc., springs of different strength must be selected to suit the patient.

The sources 'Précieuse,' 'Magdelaine,' 'Rigolette,' and 'St. Jean' are those most frequently employed for use in England.

Besides its alkaline waters, Vals likewise possesses weak sulphate of iron and arsenical springs (see p. 205).

The season is from May 15 to October 15.

The accommodation, which was unsatisfactory in former times when the waters were almost solely exported, is now sufficient.

Doctors: Chabanne, Ollier, etc.

Neuenahr, Germany (Rhenish Prussia).—Thermal alkaline waters (75°–104° F.) with a considerable amount of free carbonic acid gas. This comparatively recent spa is much resorted to. It lies in a sheltered position in the valley on both banks of the Ahr, at an altitude of about 760 feet. In the middle of summer the heat is very great, and there are not the wooded walks in the neighbourhood which might be wished for. The waters are the only simple thermal alkaline waters in Germany; they are, however, much weaker than the similar waters of Vichy. The supply is abundant, amply sufficient for drinking purposes and for use in the large roomy, recently elected bath-house. The big 'Sprudel' (temperature 104° F.) contains about 1 per mille bicarbonate of sodium, ·4 bicarbonate of magnesium, ·3 bicarbonate of calcium, and ·04 bicarbonate of iron.

There is a room where patients with catarrhal conditions of the respiratory system may inhale the finely pulverised water as at Ems; a few cases of quiescent early chronic phthisis come here during the season, but the beneficial effect is doubtful. Cases of chronic nephritis, dyspepsia with hyperacidity, and functional troubles of the nervous system (especially when allied with slight glycosuria), are treated here. The chief reputation of the spa is, however, in the treatment of troubles connected with the uric acid diathesis, and in cases of glycosuria. It hardly need be added that typical cases of diabetes in young persons are not likely to be cured here, or even to derive any permanent benefit from the treatment; the cases of so-called diabetes most likely to derive benefit being chronic cases in elderly persons, without wasting, especially when allied to gout or the uric acid diathesis. The season lasts from May to the commencement of October. Close to Neuenahr is the Apollinaris spring mentioned in the chapter on 'Table Waters.'

Access : By Cologne and Remagen in about 18 hours; Neuenahr is a station on the railway between Remagen and Altenahr.

Accommodation : Good.

Doctors : Unschuld, Grube, Lenné, Niessen, and Teschemacher.

Birresborn (Rhenish Prussia), a village with a railway station on the line from Cologne to Treves, has an alkaline gaseous spring with 2·8 per mille bicarbonate of sodium. Together with the bicarbonate of sodium it contains a little bicarbonate of magnesium and sulphate of sodium, which impart a slight laxative effect, sometimes useful in acid dyspepsia with habitual constipation. It is often used as a table water.

Fachingen, in Prussia (Province of Hesse-Nassau), on the Lahn, between Ems and Limburg, possesses an alkaline spring rich in carbonic acid gas, and of about the same strength as the water of Bilin in Bohemia. Fachingen water contains about 3·5 per mille bicarbonate of sodium. If used, therefore, as a 'table water,' it must not be expected to mix well with wine. The spring is the property of the Prussian State, and the water is only used for exportation.

Wildungen, in Germany (Principality of Waldeck).—The 'Helenenquelle' (see p. 249) may likewise be mentioned in this chapter amongst alkaline waters.

Obersalzbrunn (**Salzbrunn**) in Prussian Silesia, situated on the Salzbach at an altitude of 1,320 feet, is a station on the railway from Breslau to Halbstadt. Of its cold alkaline springs the chief one 'Oberbrunnen' contains 2·15 per mille bicarbonate of sodium, ·01 bicarbonate of lithium, ·4 per mille sulphate of sodium, and 985 per mille volumes of carbonic acid gas. The 'Mühlbrunnen' and other springs contain less bicarbonate of sodium. The 'Kronenquelle,' a spring in private hands, used for exportation, contains only ·87 per mille bicarbonate of sodium, ·01 bicarbonate of lithium, ·7 per mille bicarbonate of calcium, and ·18 per mille sulphate of sodium; the 'Wilhelmsquelle' is equally weak. The whey cure is made use of at this spa, as at so many others. Season May 1 to September 30.

Doctors: Nitsche, Pohl, Determeyer, Montag.

Radein (altitude 660 feet) in Styria possesses cold gaseous alkaline waters with 3 per mille bicarbonate of sodium, ·04 bicarbonate of lithium, ·03 sodium iodide, and ·02 sodium bromide.

Bilin in Bohemia, near Teplitz, contains the 'Biliner Sauerbrunn,' a cold alkaline spring (3 per mille carbonate of sodium), rich in carbonic acid gas. The water is used for chronic gastric catarrh, gouty complaints, and chronic bronchial catarrh. In some persons Bilin water acts as a gentle laxative, perhaps owing to the minute quantity of sulphate of sodium it contains. In gouty cases the cure may be commenced or continued at the neighbouring Teplitz. The water may be drunk pure or mixed with whey. The bath establishment possesses arrangements for hydro-therapeutic treatment. The season lasts from May 15 to the end of September.

Szinye-Lipocz, near **Eperies**, in Hungary, possesses the 'Salvator' spring, the water of which is exported. It is a weakly mineralised cold gaseous alkaline water containing, according to Prof. M. Ballo (1882), ·3 per mille bicarbonate of sodium, ·9 per mille bicarbonate of magnesium, 1·7 per mille bicarbonate of calcium, ·09 per mille borate of sodium, and ·02 per mille bicarbonate of lithium. It has been recommended in the treatment of the uric acid diathesis, and in some urinary affections, etc.

The Fellathalquellen in Carinthia (altitude 1,970 feet), about five hours distant from the railway station of Kühnsdorf, are gaseous alkaline springs, the waters of which, containing 4·3 per mille bicarbonate of sodium and 1·7 per mille bicarbonate of calcium, are exported.

Preblau (Carinthia) lies in the Lavant valley 3,280 feet above sea level. Its gaseous alkaline waters contain 2·2 per mille bicarbonate of sodium.

Al-Gyogy, a village in Transylvania, possesses three weak alkaline springs, having a temperature of about 86° F. They are used in chronic catarrhs, skin affections, chronic rheumatism, and gout.

Passugg (Switzerland) in the Grisons, an hour's drive from the railway station of Chur, lies at an altitude of 2,620 feet in the Rabiusa-schlucht. Its cold gaseous alkaline springs contain 1·9 to 5·3 per mille bicarbonate of sodium, ·6 to 1 bicarbonate of calcium, and ·01 bicarbonate of iron. The cold earthy chalybeate

'Belvedera' spring contains 2 per mille bicarbonate of calcium, and ·03 per mille bicarbonate of iron. The climate of Passugg must aid considerably in the treatment of anæmia and some kinds of dyspepsia.

Le Boulou, in France (Pyrénées-Orientales) near the Spanish frontier, possesses simple alkaline waters rich in free carbonic acid gas (temperature 59°-68° F.). The waters, which are chiefly exported for home use, are employed in dyspeptic conditions, etc., where simple alkaline waters are indicated. The 'Source du Boulou' contains 3·7 per mille bicarbonate of sodium. The 'Source Sorède' contains a small amount of iron.

Châteauneuf in France (Department of Puy-de-Dôme) lies at an altitude of 1,750 feet, on the River Sioule, about 15 miles from Riom. It possesses alkaline waters having temperatures varying from 54° to 100·5° F.; some of the colder springs, notably the 'Source Morny,' contain a considerable amount of iron (·05 per mille of the bicarbonate in the 'Source Morny').

Vidago, in the north of Portugal, has alkaline waters used in similar cases to those in which the Vichy waters are employed.

CHAPTER IX

MURIATED ALKALINE WATERS

THESE waters are employed in the same class of cases as the simple alkaline waters, especially when emaciation is feared. The common salt lessens the tendency of the sodium bicarbonate to render the urine too alkaline, and this class of waters is not so likely to produce the depressing effects which the simple alkaline waters do in some cases. In some cases the use of the bicarbonate by itself, either in the form of an ordinary medicine or in mineral waters, may produce an attack of acute gout, whereas when given together with common salt it is not so likely to do so. It may be mentioned in connection with this that common salt by itself is said by some to be of therapeutic use in the tendency to formation of uric acid calculi and gravel.

The muriated alkaline waters are useful in cases of chronic catarrh of the respiratory organs, and in these cases the climate of the spa is of great importance.

In describing the spas belonging to this group Ems and Royat will be placed as types first, and the rest in geographical order will be arranged to follow them.

Ems, Germany (Prussia, formerly Duchy of Nassau). Ems (altitude 300 ft.) is beautifully situated in the narrow valley of the Lahn on both sides of the river. It consists of a mass of hotels, villas, bath-houses with a magnificent Kursaal, etc.; these show to what an extent the place is made use of as a health resort and for pleasure. The mild climate (too hot for many in the middle of summer) contributes doubtless to the use of this spa in laryngeal and bronchial catarrhs.

There are in use nine different thermal muriated alkaline springs, of temperatures varying from 80° to 120° F.; they contain 2 per mille bicarbonate of sodium, with 1 per mille common salt, and about 500 volumes per mille carbonic acid gas; six of the springs are used internally; there is likewise a chalybeate spring (temperature 70° F.).

The baths are nicely fitted up, and furnished with apparatus for hot and cold douches; massage may be had in suitable cases. Besides the baths there are rooms for gargling and for inhaling the finely pulverised water, simple or medicated with Peruvian balsam, oil of pine, etc. These can be used in the treatment of patients for catarrhal conditions of the larynx and bronchi. Bronchitis in gouty persons is especially treated at Ems. In cases of pulmonary emphysema an apparatus is frequently used for expiration into rarefied air and inspiration from air of increased density, after the method of Waldenburg. Dr. Geissé considers the employment of such an apparatus to afford a capital pulmonary exercise when used methodically once or twice a day for fifteen minutes. Compressed air chambers can also be made use of.

The waters of Ems are also used in catarrhal and gouty dyspepsia with hyperacidity, in cystitis, and in various gouty conditions, especially in thin or weak persons, where the simple alkaline waters, as of Vichy and Neuenahr, and the sulphated alkaline waters, as of Marienbad and Karlsbad, are considered too lowering.

Good results are claimed in some cases of chronic albuminuria.

The baths are employed in cases of leucorrhœa, catarrhal conditions of the uterus and cervix uteri, and in nervous cases of dysmenorrhœa. In some cases of sterility Dr. Geissé recommends the employment of a vaginal douche whilst the patient is in the bath. As a douche tube he prefers one of metal, since the perforations can be made finer and more numerous in the metal tubes than in tubes of gum elastic and other materials.

During hot weather the shady walks on the hillsides are very agreeable for those who are able to take them. A funicular railway has recently been made up the Malberg on the south of the town, and patients may thus quickly reach the refreshing air of the woods on the summit, which is about 1,000 feet above the town.

Access: By train in 16 to 19 hours *viâ* Cologne, Coblenz, and Niederlahnstein.

Accommodation: Very good.

Doctors: Geissé, Orth, von Ibell, etc.

Royat, France (Department of Puy-de-Dôme).—Royat is picturesquely situated in the Auvergne Mountains on the lower slopes of the Puy-de-Dôme, at an altitude of 1,480 feet. It stands in the valley of the Tiretaine stream at the entrance of a deep ravine, and is about one and a quarter mile distant from Clermont-Ferrand. On account of its thermal muriated alkaline springs it has been called the French Ems.

The four springs are the Grande Source or Source Eugénie, the Source César, the Source Saint-Mart, and the Source Saint-Victor. Their temperatures are 68° F. to 95° F. The Source Eugénie, which is the warmest and most highly mineralised, contains 1·7 per mille common salt, 1·1 per mille bicarbonate of sodium, 1 per mille bicarbonate of lime, ·04 per mille bicarbonate of iron, ·035 chloride of lithium, and a trace of arsenic. Of the other three springs the Saint-Victor contains the most iron (·056 per mille of the bicarbonate) and the most arsenic (·0045 per mille of the arseniate of sodium).

The waters of Royat are rich in carbonic acid gas, and become slightly turbid on exposure to the air. They are used for drinking, bathing, and inhalation. The carbonic acid gas is sometimes employed in the form of gas-baths. The chief bath-house of Royat is well arranged; the small one (César) has very limited accommodation, and is almost reserved for the subjects of nervous affections.

The affections treated at this spa include chronic rheumatism, gout and the uric acid diathesis, atonic dyspepsia, chronic laryngitis, and bronchitis. Chronic skin eruptions and gynæcological affections may be benefited, especially when occurring in arthritic subjects. When anæmia forms a complication, the Source Saint-Victor, on account of the iron and arsenic it contains, is generally preferred, whereas the Source Saint-Mart, with its ·035 per mille chloride of lithium, has a special reputation in gouty cases. The waters are inhaled in a finely pulverised form for diseases of the respiratory system. Douches and massage are much employed when the joints are affected.

Saint-Mart is occasionally used as a table water, and courses of this water are often taken away from the spa. Saint-Victor and César are likewise exported.

The season lasts from May 15 to September 15.

Access: Viâ Paris to Clermont-Ferrand; thence by electric tramway, train, or omnibus, to Royat. *Accommodation:* Good. *Doctors:* Petit, G. H. and J. E. Brandt, Frédet, Chauvet, Puy le Blanc, etc.

Toennistein (Rhenish Prussia) in the Brohl Valley, at an elevation of 420 feet, is an hour's distance from the railway station of Brohl. In the neighbourhood is the 'Heilbrunnen,' a gaseous muriated alkaline spring containing 2·5 per mille bicarbonate of sodium, 1·6 bicarbonate of magnesium, 1·4 common salt, ·02 bicarbonate of iron, and 1,270 per mille volumes of carbonic acid gas. Near Wassenach is also the gaseous chalybeate 'Stahlbrunnen,' containing ·08 per mille bicarbonate of iron, 1·4 bicarbonate of magnesium, and 1·2 bicarbonate of sodium. There is accommodation for visitors.

Assmannshausen, Prussia (Province of Hesse-Nassau).—This summer resort, beautifully situated on the right bank of the Rhine at the foot of the Niederwald, possesses a tepid weakly mineralised muriated alkaline spring (temperature 82·8°). The bicarbonate of lithium (·028 per mille out of a total mineralisation of only 1 per mille) in the water has been supposed to exert an especial therapeutic effect in gouty cases. Muscular rheumatism, chronic catarrhal conditions of the intestinal and respiratory organs are also treated here. It seems, however, doubtful whether the effects are not those of simple tepid water, aided by climate, change in mode of living, etc.

Weilbach in the Prussian province of Hesse-Nassau. The 'Natronlithion-Quelle' belongs to the muriated alkaline group, but is noticed with the other Weilbach springs at p. 223, amongst the sulphur waters.

Gleichenberg (Styria), 930 feet above the sea, lies in a pleasant, hilly, and well-wooded country, one and a quarter hour's drive from the railway station of Feldbach, and three hours' from Graz. The place is resorted to for its mild climate and its cold gaseous muriated alkaline waters, especially in cases of chronic (including tuberculous) affections of the respiratory organs and dyspeptic conditions. The chief springs are the Constantin-Quelle (3·6 per mille bicarbonate of sodium, 1·8 common salt, 1,340 per mille volumes of carbonic acid gas), and the less gaseous Emma-Quelle. In the neighbourhood is the gaseous chalybeate Klausen-Quelle, sometimes used for anæmic patients and at the end of the treatment. The alkaline chalybeate Johannis-Brunnen, situated at a distance of about one and a half hours, is used on account of its agreeable refreshing taste. Inhalations of the pulverised water and hydro-therapeutic procedures are frequently resorted to as aids in the treatment at Gleichenberg, which enjoys a great reputation throughout the Austro-Hungarian empire and in South Germany.

Season from May to September. The accommodation is satisfactory.

Doctors: Clar, Fischer, Hoenigsberg, Kuntze, etc.

Luhatschowitz in Moravia, a quarter of an hour's distance from the village of that name, and one and a half hour's drive from the railway station of Ungarisch-Brod, lies in a pleasant valley of the Carpathians, 1,600 feet above sea-level. It possesses cold gaseous muriated alkaline waters, containing iodine and bromine, and is chiefly used for drinking. According to J. Picek (1891) the Vincenz-Quelle and the three other chief springs contain 3·0 to 4·4 per mille carbonate of sodium, 2·4 to 4·5 per mille common salt, ·007 to ·012 per mille iodide of sodium, ·02 to ·045 bromide of sodium, ·37 to ·52 borate of sodium, and much carbonic acid gas. The waters are exported, and can be well used at home. They exercise a checking influence on the tendency to uric acid gravel, and are also useful in catarrhal conditions on a gouty basis. The admixture of common salt in these waters lessens the tendency of the sodium bicarbonate to render the urine too alkaline, and to exercise a depressing effect on the constitution. There are many persons in whom the use of sodium bicarbonate by itself has this depressing influence; whilst in the same persons the combination of common salt and sodium bicarbonate is much better borne.

Season, May 15 to the end of September.

Szczawnica in Galicia is situated on the northern declivity of the Carpathians, at an elevation of 1,700 feet above sea-level. It possesses cold muriated alkaline springs rich in carbonic acid gas; of these the Magdalenen-Quelle contains 8·4 per mille bicarbonate of sodium and 4·6 per mille chloride of sodium. Chronic catarrhal affections of the respiratory organs are suitable for this spa, which possesses also means for inhalation treatment, and for whey and koumis 'cures.' It is a six hours' drive from Alt-Sandeck, the nearest railway station.

Lipik in Slavonia, not far from the railway station of Pakracz, has a sheltered position in a valley, at an elevation of 500 feet. The hottest of its weak muriated alkaline waters (temperature 147° F.) contains a relatively large amount of iodine (1·9 per mille bicarbonate of sodium, ·6 common salt, ·26 iodide of sodium).

Saint-Nectaire in France (Department of Puy-de-Dôme) lies in the Auvergne Mountains, at an altitude of 2,500 feet. There are numerous muriated alkaline springs of temperatures ranging from 50° to 115° F. The 'Source du Mont-Cornadore' contains 2 per mille of both the chloride and the bicarbonate of sodium (temperature 100·5° F.). The 'Source Rocher' contains in addition ·057 per mille bicarbonate of lithium. The water of the 'Source Rouge' contains iron, the 'Source Boëtte' is the hottest of the springs (115° F.). The waters are employed for drinking, for baths and douches, and the carbonic acid gas obtained from the water is employed for 'gas-baths.' Saint-Nectaire is two hours' drive from the railway station of Coudes. The season lasts from the commencement of June to the commencement of October. The waters of Saint-Nectaire have recently been recommended by Ducrohêt in certain forms of albuminuria, especially in those of a gouty nature, and those designated 'phosphaturic albuminuria' by Robin, when as yet the fault lies rather in the general metabolism of the body than in any organic disease of the kidneys.

La Bourboule (France) in the Auvergne.—The waters have been classed in the arsenical group (see p. 200).

Pozzuoli (Italy), the ancient Puteoli, on the Bay between Naples and Baiæ, possesses thermal weak muriated alkaline waters known to the ancients, and still employed.

CHAPTER X

SULPHATED ALKALINE WATERS

THESE waters are useful in atonic constipation, associated with 'abdominal plethora,' also in cases of hæmorrhoids and disturbance of the female pelvic organs, especially when these disorders occur in large eaters and corpulent persons, in whom loss of flesh is to be desired rather than feared.

They may be serviceable also in gastric and intestinal catarrhs, especially in those who have indulged much in the luxuries of the table, in cases of catarrhal jaundice, in tendency to the formation of gall-stones, in congestion of the liver, and in enlargement of this organ resulting from fevers and malarial affections in hot climates, in uric acid gravel, and in some cases of gout and glycosuria in fat persons.

Other waters are preferable in thin and feeble individuals.

In the results obtained by the sulphated alkaline waters diet plays a most important part, especially so in cases of glycosuria and obesity.

It is claimed that chronic enlargement of the spleen occurring as a result of malaria may derive benefit from courses of the alkaline sulphated waters; the amount of benefit obtained varies in different cases.

The spas of this group will be described in the

following order: Karlsbad, Marienbad, Franzensbad, Tarasp-Schuls, Elster, etc.

Karlsbad (Carlsbad) in Bohemia. Karlsbad (altitude about 1,160 feet) is a long narrow town stretching upwards in the narrow valley of the Tepl, on both sides of the stream, from its entrance into the Eger for about two miles in a southward direction. Owing to the somewhat cramped position of the houses of the main streets of this ever-increasing spa, some of the guests prefer to live in the buildings situated higher up on the Schlossberg, etc., where the air is fresher and purer; much, moreover, has been done, and is still being done, to broaden the main thoroughfares and open out the older portion of the spa. Beautiful walks can be enjoyed in the woods covering the slopes of the valley, and a favourite walk which needs no climbing is the one in the valley higher up along the Tepl. In one or other of the cafés along this road guests frequently breakfast after drinking the water.

There are a great many mineral springs at Karlsbad, but they are remarkably similar in their solid constituents, so similar in fact that there is supposed to be some large natural reservoir in the rocks below the town from which the springs all derive their water. Hence one may really speak of a 'Karlsbad water,' which contains about 2·4 per mille sulphate of sodium, about 1·2 per mille bicarbonate of sodium, and 1 per mille common salt, with a moderate amount of carbonic acid gas. It will be unnecessary here to mention all the sixteen springs of Karlsbad water; the chief difference between the various springs lies in their temperature, the hottest, moreover, having the least amount of carbonic acid gas.

The hottest of the springs is the famous Karlsbad

L 2

Sprudel[1] (temperature 162·5° F.), a steaming fountain leaping up at short intervals in a jerky, irregular way; close by it, along the sides of the Tepl, clouds of steam arise from the ground itself. The Felsenquelle has a temperature of 138° F.; the Schloss-Brunnen of 127° F.; and the Mühlbrunnen of 124·5° F. The Spital-Brunnen in the Strangers' Hospital has the lowest temperature (95·4° F.) of the true Karlsbad waters, for the Stephanie-quelle (temperature 71·9° F.), which arises below the Schweizerhof, at some distance from the other fountains, appears to be a spring of true Karlsbad water diluted and cooled during its passage to the surface by ordinary spring water. The Dorotheen-Säuerling, which arises close to the Stephanie-Quelle, is, as the name implies, an ordinary acidulated spring, the water of which may be used as an ordinary refreshing draught or table water. The neighbouring Giesshübl and Krondorf table waters can be used at Karlsbad.

As a general rule, the hotter springs have a less laxative action than the cooler springs. If it is desirable to take the dose cold, the water may be obtained the evening before (the Sprudel, if very little carbonic acid gas be preferred), and allowed to cool at home during the night. Owing to the great number of guests, it is important that not all be told to drink from the same fountain.

During summer the usual time for drinking the waters is from half-past five to half-past eight in the morning, about a quarter of an hour being allowed after each glass (about six ounces). When, however, a comparatively large amount is taken, the daily dose may be divided into two or three portions, a second

[1] The German term 'Sprudel' is applied to any gaseous spring which arises with sufficient force to leap up from the ground.

portion being taken in the forenoon, before the mid-day meal, and occasionally a third portion in the afternoon; this is also the case when the stomach can bear very little of the water at a time. Sometimes a dose is taken cold at bed-time. In former days enormous quantities of the water used to be taken, but now, as a general rule, the dose varies between two and six glasses (of about six ounces each), and in some cases, as for example in cases of chronic diarrhœa, the doctors begin with very small doses, such as half a glass (about three ounces), and even less.

Amongst the conditions for which the internal use of Karlsbad waters is useful, in the first place come affections of the liver, including catarrhal jaundice, frequent attacks of biliary colic, early stages of alcoholic cirrhosis, etc., enlargements of the liver in great eaters (sometimes a part of general adiposity). Then come cases of habitual constipation and hæmorrhoidal conditions in robust people; some cases of chronic gastric or intestinal catarrh with or without diarrhœa; some cases of dyspepsia apparently without organic alteration in the alimentary tract; the uric acid diathesis; chronic glycosuria in fat people, and general adiposity, which is often combined with a weakly acting heart. It is also maintained that the lesser degrees of chronic malarial enlargement of the spleen are benefited by a course of the waters. Persons with periodic or frequently recurring headaches connected with abdominal disorders are likewise treated at Karlsbad, and often with great benefit. As a general rule very feeble patients are unsuitable subjects for Karlsbad treatment.

By no means all the patients who come to Karlsbad to drink the waters require a course of the mineral baths in addition. For cases, however, in which baths are indicated, Karlsbad is well provided; it contains six bath-houses, of which the 'Kaiserbad,' just erected by the town, is the most complete, and is one of the most magnificent bath-houses in Europe. In addition to ordinary and mineral water baths there are arrangements for 'moor baths,' as at Franzensbad, the peat used for these baths being obtained from a part of the Franzensbad moor which belongs to Karlsbad. There are likewise arrangements for douches, hot-air and vapour baths, massage, and Swedish gymnastics.

In conducting the course of treatment, the spa doctor, of course, considers each case for itself with reference to the nature of the affection, the constitution of the patient, and his previous habits. In a great number of cases, such as those of glycosuria, catarrh of the stomach and intestines, obesity, etc., the regulation of the diet is of extreme importance. Formerly there was a special 'cure diet,' to which the patient was supposed to confine himself as a matter of course; thus all acid things were supposed to be antagonistic to the proper action of the Karlsbad waters; in no cases was butter allowed; and 'Sprudelsuppe,' a soup made with Karlsbad water, formed the chief part of the evening meal. All this is now much modified: the resident doctors regulate the diet, according to ordinary indications to suit the individual patient, and with due regard to his previous habits. The absence of table d'hôte at the hotels assists the patient greatly in following the doctors' orders, though it must be admitted that the general provision of midday 'couverts' (*i.e.* dinners at fixed prices) must sometimes afford temptation to neglect precise instructions as to diet.

The average daily programme of ordinary 'Kurgäste' at Karlsbad may be shortly sketched somewhat as follows:

Rising early to drink the waters at the fountain; in the interval between the glasses promenading to the sound of 'Kur-Musik;' then walking to some café, often to one of those beyond the town along the Marienbader-strasse, and taking breakfast (at about 9 A.M.). This consists of coffee or tea, with rolls and perhaps boiled eggs or ham; a curious habit prevailing at Karlsbad is that the guests, after taking the waters, often buy their rolls direct from a baker, and carry them to the place where they breakfast. At about one o'clock the chief meal is taken, then coffee or tea at about four, and a light supper in the evening. Those who have been ordered a course of baths mostly take them in the forenoon; promenades, listening to bands and concerts, or walks and excursions for those who are advised to take more active exercise, occupy the remaining intervals in the day. The old idea that *all* the invalids must take a large amount of walking exercise is now recognised as entirely erroneous.

The season lasts from the middle of April to the end of September, but guests are also received at other times of the year, though of course most of the hotels would then be shut. An 'after-cure' should always follow the course at Karlsbad (see p. 43).

Access: By various routes in about thirty-one hours: either by Cologne, Würzburg and Nürnberg or Bamberg; or by Cologne, Leipzig, Dresden, and Komotau; or by Paris, Stuttgart, and Nürnberg.

Accommodation: Good, but in the height of the season rooms should be secured beforehand.

Doctors: Von Hochberger, J. Mayer, Neubauer, Kraus, senior and junior, London, Hertzka, Kallay, Lebovici, etc.

Marienbad (Bohemia).—This now much-frequented spa is beautifully situated (at an altitude of about 1,980 feet) in a rather open valley, and sheltered by an almost complete circle of hills, amongst the pine forests on which beautiful walks may be enjoyed.

The chief of the springs are the Kreuz-Brunnen and the Ferdinands-Brunnen, which are sulphated alkaline springs resembling those of Karlsbad, but are cold instead of hot, richer in the sulphate, bicarbonate, and chloride of sodium, and in carbonic acid gas, and containing respectively ·048 and ·084 per mille bicarbonate of iron. The Kreuz-Brunnen contains about 4·9 per mille sulphate of sodium, 1·6 per mille bicarbonate of sodium, and 1·7 per mille common salt, the sulphate of sodium being about double the amount in the Karlsbad springs. The Ferdinands-Brunnen resembles the Kreuz-Brunnen, but is richer in the above-mentioned saline constituents and in carbonic acid gas (5 per mille sulphate of sodium, 1·8 per mille bicarbonate of sodium, and 2 per cent. chloride of sodium). The Waldquelle and the Alexandrinenquelle, situated respectively at the north and south ends of the town, are more weakly mineralised, but distinguished for their relatively larger amount of bicarbonate of sodium and of carbonic acid gas. (The Waldquelle has 1·4 per mille bicarbonate of sodium to one per mille sulphate of sodium.) The Ambrosius-Brunnen and the Karolinen-Brunnen are chalybeate springs, the first being much the stronger, and being said to contain as much as ·166 per mille bicarbonate of iron. The Rudolfsquelle is an alkaline earthy spring, which may be compared to the Helenenquelle at Wildungen. All the Marienbad springs are cold.

From the variety of springs it may be seen that different classes of cases can be treated at Marienbad. Patients with vesical catarrh and urinary complaints may be benefited, as at Wildungen, etc., by drinking the water of the Rudolfsquelle, and observing the proper precautions as to diet. The Waldquelle is used as an aerated alkaline spring in chronic catarrh of the respiratory organs. Anæmic patients may drink the chalybeate waters of the Ambrosius-Brunnen, if it does not interfere with their digestion.

On the whole, however, the main class of patients coming to Marienbad are those likely to be benefited by a course of sulphated alkaline waters—namely, full-blooded and stout people who have led a sedentary life and fed largely; this class of patients, suffering from dyspepsia, the uric acid diathesis, chronic constipation, hæmorrhoids, or chronic catarrh of the large intestine, or affected with general obesity, possibly with enlargement of the liver and a weakly acting heart, may often be benefited by a course of waters at Marienbad. In such cases the Kreuz-Brunnen or Ferdinands-Brunnen are mostly employed. Furthermore, hepatic troubles, such as catarrhal jaundice, gall-stones, and incipient cirrhosis, may be treated as at Karlsbad; so also chronic glycosuria in obese or fairly well-nourished persons. In the glycosuric and hepatic cases, as well as in many others, it is better to have the waters warmed, by which process they become very much like the waters of Karlsbad.

At Marienbad, as at Karlsbad, treatment by baths takes a secondary place, but there are four bath-houses, and all sorts of baths may be obtained. An eighth Marienbad spring, the Marienquelle, poor in solid constituents, but rich in carbonic acid gas, is used for mineral-water baths; water from the Ferdinands-Brunnen and the chalybeate Ambrosius and Caroline springs are also used. 'Moor baths' (general and local) are given as at Franzensbad, and may be used for a variety of chronic gynæcological affections, or in some cases merely as a variety of thermal bath. The ferruginous peat, used for Marienbad moor baths, is asserted to be as rich or richer in iron than the Franzensbad peat. Hot-air and vapour baths, douches, and gas baths may be obtained.

The season lasts from May to September. An 'after-cure' is always advisable after a course of waters at Marienbad (see p. 43).

Access: Similar to that for Karlsbad.

Accommodation: Very good.

Doctors: Ott, senior and junior; von Basch, Herzig, Porges, Kisch, von Heidler-Heilborn, Opitz, etc.

Franzensbad, in Bohemia.—Franzensbad, near Eger, founded by the Emperor Francis II. in 1793, is situated in a flat part of the country at an elevation of about 1,500 feet above the sea-level. The moorlands, whence is derived the peat used for the famous 'moor-baths' of Franzensbad, immediately adjoin the town.

There are twelve different mineral springs, as well as a simple acidulated spring, which resembles ordinary effervescent 'table-waters.' These twelve springs are all cold and rich in carbonic acid gas, but in their solid mineral constituents differ considerably. The Salzquelle, Franzensquelle, Wiesenquelle, and Kalte Sprudel, all used for drinking, are sulphated alkaline springs, whose waters contain 2·7 to 3·5 per mille sulphate of sodium, ·67 to 1·1 per mille carbonate of sodium, traces of other carbonates, about 1·2 per mille common salt, with ·009 to ·030 per mille of the carbonate of iron. Of these springs the Salzquelle contains the least amount of iron (only ·009 per mille of the carbonate), and is, therefore, the one which most resembles the waters of Karlsbad; this resemblance can be increased by warming its waters to the temperature of one of the Karlsbad springs. The Neuquelle is similar, except that it is now said to contain much more iron than the others. The Stahlquelle contains about ·079 per mille bicarbonate of iron, and less of the other salts (1·6 per mille sulphate of sodium, with a total mineralisation of 3·1 per mille), so that it may fairly be ranked as a strong chalybeate spring.

Franzensbad possesses four well-provided bath-houses, at which the three principal kinds of baths employed are the following: (1) The 'Stahlbäder.' This term is somewhat confusing. By it the mineral-water baths are meant, in which the warming process is so arranged (method of Schwarz) as to occasion the least possible escape of carbonic acid gas. The heating in these baths is effected by a steam chamber or steam pipes at the bottom of the bath. (2) The 'Luisenbäder' or 'Mineralbäder.' These are the ordinary mineral-water baths, in which steam is passed through the mineral water to heat it (method of Pfriem), necessitating the escape of the greater amount of carbonic acid gas. Thus, by the loss of the gas, the 'Luisenbäder' are rendered less stimulating than the 'Stahlbäder.' (3) The 'Moorbäder,' for which Franzensbad has attained such a notoriety. The peat used for these baths is obtained from moorland in the immediate neighbourhood of the town, and the supply is so plentiful that the peat used for one bath need never be used a second time. As much as 25 per cent. by weight of the disintegrated peat, when ready for use, consists of substances soluble in water, and it is said to be extraordinarily rich in sulphate of iron. The usual temperature at which the 'moor-baths' are given is 89·5° to 95° F.; they act as a huge poultice to the abdomen and lower limbs, and should not cover the upper part of the chest. Local peat baths, still more resembling poultices, are likewise given.

Besides the above-mentioned three chief kinds of baths, general and local baths of carbonic acid gas are likewise sometimes employed at Franzensbad. In the general gas-bath the patient, clad in a light bathing-dress, sits or stands in a sunken space, into the bottom of which a pipe leads, conveying carbonic acid collected from the mineral waters. An overflow-pipe carries off the carbonic acid at a certain height, and thus avoids the danger of the patient inhaling it. A subjective sensation of warmth in the lower limbs and the part of the body bathed by the gas is produced, but the exact therapeutic value of these gas-baths remains doubtful.

At Franzensbad the daily dose of water is often divided into two portions, one to be taken before breakfast and one later on in the forenoon.

Owing to the differences between the various mineral springs at Franzensbad, different classes of cases can be treated at this spa. The sulphated alkaline Salzquelle can be used internally, warmed if necessary, in the same class of cases as the Karlsbad waters; the sulphated alkaline Neuquelle, Franzensquelle, etc., are useful, owing to their iron constituent, in anæmic conditions associated with constipation; whilst the purer chalybeate 'Stahlquelle' may be given in such anæmic cases as are benefited by the waters of Pyrmont, etc.

Franzensbad has, however, obtained an especial reputation as a 'ladies' spa,' and by far the majority of the 'Kurgäste' belong to the female sex. They include girls and women suffering from chlorosis and other anæmic or cachectic conditions; cases of functional nervous troubles often allied with a debilitated condition of the whole body;

dyspeptic troubles in which those Franzensbad waters most similar to the Karlsbad ones are likely to be useful; chronic rheumatic and gouty affections, when likely to be benefited by a judicious course of baths; lastly, there are the patients suffering from various affections of the pelvic organs. Of the latter class some are anæmic and likely to be benefited by drinking the iron waters, others are benefited by drinking from the more laxative springs; the 'Stahlbäder' or the 'Luisenbäder' exert a favourable effect in leucorrhœa and catarrhal conditions of the pelvic organs, and the 'Moorbäder' are said to promote the absorption of old pelvic exudations. Broadly speaking, the 'moor-baths' are contra-indicated in diseases with a tendency to acute exacerbations, in diseases of the heart and blood-vessels, in tendency to hæmorrhage from various organs, and during pregnancy and generally during menstruation.

The season at Franzensbad lasts from May to the end of September. An 'after-cure' following treatment at Franzensbad is always to be desired (see p. 43).

Access: Similar to that for Karlsbad.

Accommodation: Good.

Doctors: Klein, Sommer, Straschnow, Fellner, Steinschneider, Egger, &c.

Tarasp-Schuls (Tarasp-Schuls-Vulpera), Switzerland (Canton Grisons).

The Kurhaus of Tarasp is situated in the Lower Endadine Valley, on the river Inn at an altitude of 3,870 feet; the positions of the neighbouring villages of Ober-Schuls (altitude 4,060 feet) and Vulpera (altitude 4,160 feet) are rather preferable, especially that of the latter. Its mineral waters have been known since the sixteenth century, but it is only in comparatively recent times that they have been duly appreciated. There are eight cold mineral springs which are used; four yield a sulphated alkaline water, known in the neighbourhood as 'Salzwasser;' four yield gaseous chalybeate water known as 'Sauerwasser.'

Amongst the sulphated alkaline springs, the Lucius and the Emerita, the two used for drinking, contain 2·1 per mille sulphate of sodium, which is about the same amount as that of the Karlsbad springs, but they are cold and much richer in bicarbonate of sodium, common salt, and carbonic acid gas than the Karlsbad water. The amount of sodium bicarbonate is 4·8 per mille, equalling that of Vichy; of common salt 3·6 per mille; of bicarbonate of calcium 2·4 per mille, and of bicarbonate of iron ·02 per mille. The Emerita spring is not quite so rich in gas as the Lucius spring. The Ursusquelle and the Badequelle are used for baths (gaseous saline baths).

Amongst the chalybeate springs the 'Bonifacius' is the strongest, containing ·045 per mille of the bicarbonate of iron, together with 1·4 per mille bicarbonate of sodium, 2·7 per mille bicarbonate of calcium, and much free carbonic acid gas. The iron baths are heated by a tube of steam, so as to cause comparatively little of the carbonic acid to escape.

About three hours distant is the chalybeate spring of Val Sinestra (*q.v.*), rich in carbonic acid gas and containing arsenic. The Val Sinestra water can be now obtained at Schuls, freshly brought each day from the spring.

Besides the baths of the Tarasp mineral waters, Rheinfelden Soolbäder, and Battaglia mud baths may be had, if required.

The sulphated alkaline water is employed internally in chronic constipation, hæmorrhoids, dyspeptic conditions, and catarrh of the bowels when occurring in stout full-blooded persons; in cases of gall stones; in glycosuria of fat persons, etc. It may be made more to resemble the Karlsbad water by warming it before drinking; and this is especially important in cases of gall-stones and allied affections.

The action of the chalybeate waters of Tarasp in anæmic and debilitated conditions is doubtless greatly aided by the Alpine climate. The arsenic contained in the water of Val Sinestra might exert a special influence in cases of malarial cachexia.

The season lasts from June 15 to September 15. Regarding 'after-cures,' see p. 43.

Access: 6 hours by diligence over the Fluela pass from the railway station of Davos; 9 hours by diligence from the station of Landeck (for those coming from the north-east).

Acommodation: Good.

Doctors: Leva, Dorta, and Vogelsang.

Elster (Bad Elster) in the Kingdom of Saxony.—Elster (altitude 1,550 feet), situated near the Bohemian frontier in a valley protected from the east wind, possesses waters of two classes.

The 'Salzquelle' belongs to the sulphated alkaline group (5·2 per mille sulphate of sodium, 1·6 per mille bicarbonate of sodium, ·8 per mille common salt, ·06 per mille bicarbonate of iron, and much carbonic acid gas), and is said by Pollach and Flechsig to rank between the 'Kreuz-Brunnen' and the 'Ferdinands-Brunnen' of Marienbad; the indications for treatment are the same for this spring as for those of Marienbad.

The other springs of Elster are compound chalybeate, and of these the one used for drinking, 'Marienquelle,' contains ·06 per mille bicarbonate of iron, ·7 per mille bicarbonate of sodium, 1·8 per mille common salt, 2·9 per mille sulphate of sodium, and much carbonic acid gas. Owing to the admixture of laxative saline constituents the Elster chalybeate waters differ in their action somewhat from the pure chalybeate waters of Schwalbach, and resemble rather the compound chalybeate waters of Franzensbad. They may be used especially in those cases of anæmia in which there is a tendency to constipation. The 'iron baths' of Elster resemble other gaseous baths in their action. Ferruginous 'moor-baths' are likewise made use of. The season lasts from May 15 to September 20.

Access: Elster is a station on the railway between Reichenbach and Eger.

Accommodation: Satisfactory.

Doctors: Peters, Pässler, Hahn, Helmkampff, Bechler, etc.

Bertrich (Rhenish Prussia).—This spa, situated in the Uesbachthal between Treves and Coblenz, one hour from the railway station of Bullay, possesses tepid waters (91° F.), containing the sulphate, bicarbonate, and chloride of sodium, with free carbonic gas. The mineral waters of Bertrich are in their constituents similar to those of Karlsbad (*q.v.*), but only about one-third as strong. In their internal action they are therefore much weaker than the Karlsbad waters, and approach the indifferent thermal group. They are used internally in some cases of dyspepsia, gouty and rheumatic complaints, and the uric acid diathesis. The tepid baths exercise a soothing effect in irritable neuroses. The season lasts from May 1 to the end of September.

Rohitsch, Rohitsch Sauerbrunn, or Heiligen-Kreuzbad (Styria), three hours from Cilli and one and a quarter hour's drive from the railway station of Pöltschach, possesses a mild climate and beautiful situation 730 feet above sea level. Its cold gaseous springs are rather weak members of the alkaline sulphated group. The Tempel-Brunnen and the Styria-Brunnen are the only ones employed for drinking; the former contains 3 per mille sulphate of sodium, and about 1 per mille each of the bicarbonates of sodium and magnesium; the Styria-Brunnen is similar, but contains much more bicarbonate of magnesium (4·5 per mille). The amount of common salt in the Rohitsch waters is under 1 per mille.

Though much weaker in sulphates than the cold alkaline sulphated springs of Marienbad, those of Rohitsch are found useful in cases of dyspepsia associated with constipation, gastric and intestinal catarrh, etc. The season is May 1 to the middle of October.

CHAPTER XI

SULPHATED AND MURIATED-SULPHATED WATERS

The sulphated waters (see p. 24) are much employed for their simple aperient action in constipation and dyspepsia allied with constipation, especially in stout and full-blooded persons. Since the stronger waters of this class are chiefly used as occasional aperients, they are exported and taken in the patient's home more frequently than at the spring itself. Many patients prefer taking natural purgative waters to other aperients.

Amongst the best known are Franz-Joseph, Hunyadi-Janos, Aesculap, Apenta, and the many other 'Hungarian bitter waters;' the Rubinat and Condal waters of Rubinat in Spain; Birmensdorf in Canton Aargau, near Baden in Switzerland; Püllna, Sedlitz,[1] and Saidschütz in Bohemia; Galthof, near Brünn, in Moravia; *Eau Verte* of Montmirail in the Department of Vaucluse in France. Some of these waters are very strong, that of Gran in Hungary containing 4½ per cent. sulphate of magnesium, and those of Rubinat and Carabana in Spain containing about 10 per cent. sulphate of sodium. Of those mentioned above the weakest are 'Galthofer,' containing 7·4 per mille of sulphate of magnesium, with 4·9 of sulphate

[1] The 'Sedlitz powders' of apothecaries are made with tartaric acid, and, therefore, of course do not imitate the constituents of natural Sedlitz water.

of sodium, and Sedlitz, containing 13·5 per mille sulphate of magnesium.

Little need be said of the English sulphated waters, which have at one time or another been employed. Amongst them are those of Victoria Spa in Warwickshire, Purton Spa in Wiltshire, Cherry Rock in Gloucestershire, Scarborough in Yorkshire, and the original spring (practically no longer used) at Epsom. The sulphated waters near London, of Kilburn, Sydenham Wells, Beulah, Streatham, Barnet,[1] and Northaw, were all at one time much employed, those of Streatham till quite recently.

We shall now proceed to the *muriated-sulphated* springs. The waters of this group contain a considerable proportion of common salt, sufficient to modify the action of the sulphates. FRIEDRICHSHALL in Saxe-Meiningen possesses a bitter water, containing a considerable amount of common salt (24 per mille) and chloride of magnesium (12 per mille), together with sulphate of sodium (18 per mille). The figures given in brackets are those of Prof. Oscar Liebreich, but according to Justus von Liebig (1846) and Bernhard Fischer (1894) the proportion of solid constituents is less. The common salt in this water is supposed to enable it to be taken for a longer period than other bitter waters without disturbing the digestion.

Brides-Salins (France, Savoy) includes the neighbouring spas of BRIDES-LES-BAINS and SALINS-MOUTIERS. Brides-les-Bains is situated in the valley of the Doron

[1] In Charles II.'s reign such waters were apparently taken at the wells early in the morning, as laxative saline waters are now usually taken at foreign spas. Pepys, in his diary, mentions how on a very cold morning, August 11, 1667, at seven o'clock, he found many people drinking the waters at Barnet Wells.

at an elevation of about 1,860 feet above the sea. It is about three miles from the railway station of Moutiers, and two miles from Salins-Moutiers. The rather weak muriated-sulphated springs have a temperature of 95° F., and contain about 1·9 per mille common salt, about 1·2 per mille sulphate of sodium, ·5 per mille sulphate of magnesium, and about 1·7 per mille sulphate of calcium. The waters in small doses have a tonic 'eupeptic' action according to Dr. Delastre, but in larger doses have a laxative action. They are used internally in chronic constipation, dyspepsia with constipation, and the uric acid diathesis, and have lately been recommended by Delastre in certain gouty cases of albuminuria, and in forms of albuminuria, such as that termed 'phosphaturic albuminuria,' by Robin, depending rather on a vice of the general nutrition than on any organic change in the kidneys. The daily dose required to produce a laxative effect varies much in different individuals; in some cases it is necessary to add a dose of the Brides salts to produce it. The common salt in these waters renders their action less debilitating than that of ordinary sulphated waters. The Brides waters are also used for baths, but the visitors at Brides often bathe at the neighbouring Salins-Moutiers.

Salins-Moutiers (altitude 1,610 feet) lies in the valley of the Doron, about a mile from the railway station of Moutiers. Its muriated waters (temperature 95°) contain 1 per cent. common salt, and a small amount of the sulphates of calcium, sodium, and magnesium. They are rich in carbonic acid gas, and are used chiefly for baths in scrofulous and rickety children, convalescent and delicate persons, etc. The season lasts from June 1 to October 1.

Accommodation: Good.

Doctors: Delastre, Desprez, Gonthier, Laissus, Philbert.

Saint-Gervais (France, Department of Haute-Savoie).—The spa of Saint-Gervais lies in a gorge about 2,000 feet above the sea level, near the much higher village of Saint-Gervais, and in the neighbourhood of the grand Alpine scenery of Chamounix. It possesses three thermal muriated-sulphated springs, the Source de Mey (108° F.), the Source de Gontard (102° F.), and the Source du Torrent (102° F.) According to Willm's analysis of 1889, they contain 1·7 per mille sulphate of sodium, 1·7 per mille chloride of sodium, and ·9 per mille sulphate of calcium. The Source du Torrent is the only one which contains sulphuretted hydrogen.

These waters (slightly laxative in large doses) are employed in cutaneous affections, chronic rheumatism, dyspepsia with chronic constipation, etc.

According to Égasse and Guyenot the so-called 'chalybeate spring' of Saint-Gervais no longer contains any iron. The season is from June 1 to October 1. Saint-Gervais is reached from the railway station of Cluses (15 miles) in a little over two hours by carriage. The bath-establishment has been rebuilt since the disaster of 1892.

Doctor: A. Wisard.

Leamington (England, Warwickshire).—Leamington is situated in a very beautiful and historically interesting part of England. It possesses muriated-sulphated waters, having a minute quantity of carbonate of iron. The pump-room stands in the lower part of the town, on the right bank of the river Leam. On the other side of the road is the Jephson public garden, named after Dr. Jephson, to whose management and reputation at the early part of the present century the development of the spa is largely due. The garden to some extent rivals the 'Kurgarten' of Continental spas.

According to Prof. Brazier's analysis the water of the 'Public Fount' contains about 8·5 per mille common salt, 1·2 per mille sulphate of sodium, 2·0 per mille sulphate of calcium, and ·87 per mille sulphate of magnesium. The water of the 'Aylesford Well' contains slightly more sulphate of sodium.

Leamington is resorted to by those who suffer from hepatic troubles after long residence in hot climates; by those who have too freely indulged in the pleasures of eating and drinking; by those suffering from chronic gouty and rheumatic affections. The presence of the sulphates of sodium and magnesium imparts a slightly aperient action to the waters (if as much as a pint is taken), which is useful in the preceding classes of cases and in chlorosis, when combined with iron medicines. The diet is regulated by the local doctors.

In suitable cases, besides the internal treatment and baths, massage and various hydro-therapeutic appliances are employed, which are specially useful in the treatment of old adhesions about joints, the results of rheumatism or injuries. The 'Nauheim' treatment for affections of the heart has been practised lately at Leamington.

Access: From Euston Station (London) in three hours.

Accommodation: Good.

Doctors: F. Thorne, Thursfield, Eardley Wilmot, F. W. Smith, Haynes.

Cheltenham (England, Gloucestershire).—Cheltenham is a flat town, lying in the Severn Valley and sheltered from east winds. It possesses muriated-sulphated and chalybeate waters. The chalybeate waters are represented by the 'Cambray Chalybeate Spring,' which, according to an old analysis, is said to contain as much as ·1 per mille of the carbonate of iron. According to Prof. T. E. Thorpe's analysis of 1893 the 'Lansdowne Terrace Well' contains about 5·6 per mille common salt, 2·2 per mille sulphate of sodium, and ·7 per mille sulphate of magnesium. The three 'Pittville Wells' have no sulphate of magnesium, whilst the 'Chadnor Villa Well' and the 'Cottage Well' have 1·7 and 1·8 per mille sulphate of magnesium, but only ·4 to ·6 per mille common salt.

Cheltenham has been made a special resort by those who have suffered from a prolonged residence in hot climates and those who suffer from gouty affections. Owing to the competition of foreign spas the town has lately not been nearly so much resorted to as it was at the commencement of the century. It is proposed to erect a new bathing establishment with modern hydro-therapeutic appliances.

Access: From London (Paddington station) in about 3½ hours.

Accommodation: Good.

Doctors: Wilson, Ward-Humphreys, Bennett, etc.

Melksham (England, Wiltshire, 13 miles east of Bath), like Cheltenham and Leamington, possesses muriated-sulphated waters. It likewise possesses a chalybeate spring.

Grenzach (Grand-Duchy of Baden) lies at the foot of the Rebberg, at an elevation of 920 feet above sea level, about four miles by railway from Bâle. It possesses a cold muriated-sulphated water, containing earthy salts, and poor in free carbonic acid gas (3·2 per mille sulphate of sodium, 1·9 common salt, 1·1 sulphate of calcium, ·7 bicarbonate of calcium), which is made use of in the 'Emilienbad' for drinking and bathing in cases of dyspepsia, gall-stones, hæmorrhoids, etc.

Karlsbad near **Mergentheim**, in Würtemberg, possesses the Karlsquelle, a cold muriated sulphated spring, rich in carbonic acid gas. This water contains 13·3 per mille common salt, 3·7 per mille sulphate of sodium, and 2·5 per mille sulphate of magnesium. It is used in some cases of chronic constipation, chronic catarrh of the stomach and intestines, etc. The spa contains a bath establishment and arrangements for the reception of guests.

Salzerbad (Lower Austria) is situated at an altitude of 2,000 feet, near the railway station of Hainfeld. It possesses muriated-sulphated waters and arrangements or baths.

CHAPTER XII

IRON OR CHALYBEATE WATERS

IRON waters (see also p. 24) are useful in various forms of anæmia, especially those due to previous illness or actual loss of blood. Those containing carbonate of iron with carbonic acid gas are more likely to be well borne by the stomach than those containing the more active protosulphate and persulphate. A tendency to constipation, when simply due to debility, is no contra-indication; but in cases when there is dyspepsia with intestinal catarrh, or a tendency to hepatic disorder, their use is better preceded or accompanied by that of muriated or sulphated-alkaline waters, or aperient drugs. In this way they have often to be used in the ordinary chlorosis of girls, as well as in malarial cachexia, or cachexia from residence in tropical climates. Iron waters are contra-indicated in feverish conditions and in severe disturbance of the digestive organs.

Owing to the improvement in the quality of the blood and in the general nutrition of the body, functional nervous affections, neuralgias, sterility, and impotency, when dependent on general debility, are often remedied by the use of these waters.

'Iron baths,' such as those of Spa and Schwalbach, owe their principal effect to the mechanical stimulation of the skin by the bubbles of carbonic acid gas. In baths containing sulphate of iron a useful astringent effect may be exerted on the vagina of women with leucorrhœa, and on the skin of persons with great tendency to sweating.

Amongst chalybeate spas, Spa, Schwalbach, and St. Moritz, some of the best known of this group, will be described first. The other ones will follow in the political geographical order previously made use of, excepting the sulphate of iron waters and some of the less important of the carbonate of iron waters, which will be mentioned at the end of the chapter.

Spa (Belgium, Province of Liège).—Spa was such a noted and fashionable health resort in the seventeenth and eighteenth centuries that it has given its name as a generic term to all places resorted to on account of the therapeutic virtues of their waters. Unlike many other spas, whose reputation was formerly great, the original spa has maintained its fame as a health resort, though at one time it appeared to be mainly frequented for its gambling tables and its social amusements, both of which still continue to attract many. The town is situated in a sheltered valley at an elevation of about 1,000 feet above the sea-level; it is beautifully laid out with promenades and avenues, and is surrounded by wooded hills with delightful shady walks, where the fresh air and charming views encourage exercise.

The waters may be classed with those of Schwalbach, etc., as comparatively pure chalybeate, containing a considerable amount of the bicarbonate of iron and a large amount of free carbonic acid gas. The latter makes them pleasant to most people, in spite of a faint trace of sulphuretted hydrogen. The springs chiefly used for drinking are the two situated within the town, of which the 'Pouhon de Pierre le Grand' contains about ·1 per mille[1] bicarbonate of iron, whilst the 'Pouhon du Prince de Condé' is stated to contain more. The water is cold, and in order that the carbonic acid gas shall not escape it is not warmed, as at St. Moritz, but it is recommended to suck in the water through a glass tube, so as to insure that the stomach be not disagreeably chilled by the sudden swallowing of a glass of cold water. It is only as a precaution against this that the glass tube is of any use, and is not required to protect the teeth, as is commonly supposed.

In former times enormous quantities of the water were drunk, but now the amount recommended rarely exceeds thirty ounces daily, and smaller quantities are taken at the commencement of the 'cure.' The best time for taking the waters, in the majority of cases, is in the early morning on an empty stomach, between six and eight o'clock; and at this time, in the freshness of the morning in the beautiful 'Promenade de Sept Heures,' the patients can really enjoy the stroll between their glasses. Many exceptions are, however, made to this rule. Part of the daily dose may be taken in the forenoon before lunch, or in the afternoon before dinner. Excursions may be made from Spa, and the water of one of the beautifully situated springs in the neighbourhood may be drunk instead of that of the centrally situated springs. At present, however, the neighbouring springs of Sauvenière, Géronstère, Tonnelet, and Barisart are probably more visited by tourists than by actual patients. Very weak patients may have a small glass of milk or coffee or a biscuit

[1] The different analyses appear to have given very different results. The amount of bicarbonate of iron in the 'Pouhon de Pierre le Grand' is variously given as ·07 to ·19 per mille; the amount in the 'Pouhon du Prince de Condé' is stated to be ·27 per mille.

before drinking the waters, or may take the waters before lunch or dinner instead of in the early morning. Only in rare cases or in bad weather should the waters be taken in the patient's apartment.

The affections chiefly treated at Spa are chlorosis and anæmia in women; menorrhagia, a disposition to abortion and other conditions dependent on a feeble general state of health; atonic dyspepsia or a simple tendency to diarrhœa in anæmic persons; anæmic and debilitated conditions resulting from prolonged residence in the East, and from past disease of various kinds. In leucorrhœa and relaxed conditions of the female pelvic organs the baths are of great use, and treatment at Spa has a reputation in cases of sterility when dependent on poor general health, and local catarrhal conditions of the uterus.

The bathing establishment at Spa is one of the finest and best arranged ones in Europe, and the bath rooms have the advantage of being large and airy. The establishment is supplied with mineral water by a special spring. The iron baths are considered to act chiefly by the large amount of carbonic acid gas, and the mechanical effect which the bubbles of this gas (as in other gaseous 'iron-baths') exert on the nerve-endings in the skin. By an outer chamber, at the bottom of the bath, the waters can be heated to any temperature required, without driving out too much of the gas. Much valued by the doctors at Spa is the hip bath of warm running water, in which the water is heated by a special apparatus before entering the bath; this is much used in leucorrhœa and female pelvic disorders; the vagina can be kept open by a wire speculum, which the patient can herself introduce, to facilitate the contact of the moving water with every part of the vagina during the bath. Cold water douches are likewise much used at Spa, and hot moor baths, similar to those at Franzensbad and other German spas, can be given in suitable cases, but the latter seem to have too fatiguing an effect on the majority of the weak class of patients who come to Spa for treatment.

The ordinary iron baths are much used in chlorosis and anæmia, and aid the beneficial effect of taking the waters internally. Sometimes the waters are much better borne internally after a preliminary course of baths. Dr. Scheuer recommends that, when possible, the bath should be taken early in the morning, the patient rising at six and taking the bath before drinking the waters. The baths are given hot for patients having neuralgic pains, aching in the back, or a tendency to rheumatism; afterwards they may be taken cooler, and finally the more stimulating cold-water treatment may be substituted; the cold wet sheet may sometimes be used as a transitional treatment before commencing the cold douches. Occasionally the skin does not react to the carbonic acid stimulation of the iron baths, and the more powerful treatment by cold water douches is to be preferred from the commencement; in a few cases the ordinary iron baths produce too much irritation. Dr. Scheuer does not advise women to suspend the internal use of the waters during their menstrual periods, but on the whole considers temporary discontinuation of the baths advisable.

Constipation, when produced by drinking the waters, must be rectified by a dose of imported Hungarian 'bitter water,' or some other laxative. At the commencement of the cure, neuralgic pains may be increased, but as this exacerbation soon passes away, recourse may be had to temporary opiate treatment by hypodermics or otherwise. In some cases ordinary pharmaceutical treatment may be combined with that by the Spa waters; thus iodide of potassium may be given when the anæmia is partly due to old syphilis; quinine and arsenic in malarial cases, etc. Regarding contra-indications it may be stated as a general rule that patients inclined to corpulence or to 'abdominal plethora,' those who are 'full-blooded,' and those with considerable arterial degeneration or heart disease, are unsuitable for treatment at Spa. Season, from May to October.

Access: In about thirteen hours from London by Calais or Ostend and Brussels, changing trains at Pepinster.

Accommodation: Very good.

Doctors: De Damseaux, Scheuer, Schaltin, Cafferata, etc.

Schwalbach, Germany (Prussian Province of Hesse-Nassau).—Schwalbach, officially called LANGENSCHWALBACH, to distinguish it from other Schwalbachs, lies at an altitude of about 950 feet in a branch of the Aar valley, in the northern part of the Taunus range. It is a long narrow town, the upper, south-western portion of which is more modern and comfortable, and constitutes the spa proper.

The waters are cold, strong, fairly pure chalybeate, with excess of free carbonic acid gas, similar to the waters of Spa, but with no trace of sulphuretted hydrogen. Of the different springs, the Stahlbrunnen and the Weinbrunnen are the most used internally. The Stahlbrunnen contains most iron; it is said to contain ·08 per mille bicarbonate of iron to the ·06 per mille of the Weinbrunnen. Minute quantities of bicarbonate of manganesium likewise occur in the waters. The 'Lindenbrunnen,' one of the springs used for baths, contains only ·01 per mille bicarbonate of iron, and may be classed with simple acidulated waters. Schwalbach is a very popular spa, especially amongst English and Americans.

The 'iron baths' owe their effect, as at Spa, to the mechanical stimulation of the skin by bubbles of carbonic acid gas. The baths are made of copper, so that the water can be heated with steam from a chamber at the bottom, and that the least possible loss of carbonic acid gas takes place. Peat baths are likewise given; the peat, obtained in the neighbourhood, is mixed with the mineral water, and heated in wooden tubs with steam to the required temperature. The peat baths are often useful before the ordinary 'iron baths' are commenced, but patients must rest after them to avoid fatigue. The ordinary iron baths may be used somewhat cooler as the patient gets better, and reaction takes place more readily. It is expected that a new building will shortly be added to the existing baths, and will be devoted to peat baths. Massage and ordinary hydro-therapeutic treatment can be employed in suitable cases.

The affections treated at Schwalbach are chlorosis in girls and young women, anæmic conditions and prolonged convalescence in men and women, leucorrhœa and chronic inflammatory conditions of the female pelvic organs, and disorders of the digestive system when partially or wholly dependent on a general condition of anæmia or debility. In leucorrhœa vaginal douches of the mineral water are employed, as well as the baths. The best time for taking the waters is after the mineral baths in the forenoon or in the early morning before breakfast: respecting this, however, the doctors are guided much by the strength and previous habits of their patients. Sometimes the water is recommended to be taken at the midday meal with or without the addition of a white Rhine wine. Massage of the stomach may be useful in some patients to counteract the constipating action of the water.

Access: By railway *viâ* Cologne and Wiesbaden in about 21 hours; or by Cologne and Diez; or two and a half hours' drive from the station of Eltville.

Accommodation: Very good.

Doctors: Franz, Genth, Oberstadt, Boehm, Frickhöffer, etc.

St. Moritz, Switzerland (Grisons).—St. Moritz-Bad (altitude 5,800 feet), in the valley of the Upper Engadine, is situated in the level ground between the Lake of St. Moritz and the Lake of Campfer. It is here that the springs rise. St. Moritz-Dorf lies on higher ground (altitude 6,100 feet), about 1¼ mile distant from the baths. Those that drink the waters can stop at either the village or the baths of St. Moritz, but in recent years the village has acquired a special importance of its own as a climatic winter health resort for phthisical and neurasthenic patients, and the air at the village is on the whole more bracing than that in the immediate neighbourhood of the spring. Campfer is likewise a good place to stay at.

There are three different cold chalybeate springs, all rich in carbonic acid gas: the Altequelle, or Badequelle; the Neuequelle, also named Paracelsus-Quelle, in honour of Paracelsus, who in his writings mentioned the waters of St. Moritz; and, lastly, the recently discovered Surpunt-Quelle. The two first mentioned contain ·033 and ·038 per mille bicarbonate of iron, ·27 and ·18 per mille bicarbonate of sodium, and about 1·2 per mille bicarbonate of calcium; the third spring contains about the same amount of iron, less earthy material, and more free carbonic acid gas than the first two springs.

Owing to the amount of carbonic acid gas, the water of St. Moritz is pleasant to the taste; it may be taken by those who are strong enough in the morning before breakfast, or else in the forenoon about an hour before the midday meal, or in the afternoon a couple of hours after; sometimes it is taken with meals.

When compared with the springs of Schwalbach, etc., those of St. Moritz are relatively weak in iron and scarcely representative of the chalybeate group, but owing to the climatic advantages of the place they are more effective in many cases than stronger springs at lower situations. On the other hand, there are nervous, excitable patients, who are not suited for the high altitude and dryness of the air of St. Moritz; anæmic cases complicated with albuminuria likewise do not bear the climate well. In the case of feeble patients, or those with excitable hearts, it is advisable to rest some time preliminarily at an intermediate station of somewhat lower altitude, such as Churwalden, or Parpan, Savognin, or Bergün. The season for the baths of St. Moritz is June 15 to September 15.

Access: From the railway station of Chur by diligence to St. Moritz in about 13 hours.

Accommodation: Very good. During the season it is advisable to secure rooms in advance.

Doctors: Veraguth, Holland, Tidey, Berry, Hoessli, Bernhard (Samaden), etc.

Tunbridge Wells, England, Kent.—The water of Tunbridge Wells (altitude about 420 feet) belongs to the pure chalybeate class, and, according to the analysis by Dr. Y. Stevenson in 1892, contains about ·06 per mille carbonate of iron. This spa was a most fashionable resort in the last century, when Bath was at the height of its prosperity. It is now a very popular health resort, and still crowded with visitors, but only a few come to drink the waters. There are several commons in the neighbourhood, affording excellent opportunity for walks in the fresh air.

The pump-room and 'Pantiles' (an old-fashioned arcade of shops) are situated in a hollow, and one has to descend to get to them from any other part of the town. The waters are only employed internally, and do not contain the carbonic acid gas, so important in the internal and external use of the foreign chalybeate waters of Schwalbach, St. Moritz, etc.

Doubtless the climate largely contributes to the benefit derived by anæmic and enfeebled persons at Tunbridge Wells. In bad cases of chlorosis the climate should be assisted by pharmaceutical preparations, especially if the mineral water is not well digested. The chief season for Tunbridge Wells is from June to September.

Access : About an hour and a half by rail from London.

Accommodation : Good.

Doctors : Ranking, Rix, etc.

Stafford and Saltburn, which have been mentioned amongst brine baths, likewise possess chalybeate waters. So do Cheltenham and Melksham, which have been mentioned amongst the muriated-sulphated waters; Harrogate, described in the sulphur class; and Buxton, described in the indifferent thermal group.

Amongst other chalybeate waters which are known or have been known in England one may mention: Flitwick Well, near Ampthill, in Bedfordshire (whose ferric sulphate waters are sold in bottles); Sandrock, near Blackgang Chine, in the Isle of Wight (containing alum); Gilsland Spa (there is also a sulphur well here) in Cumberland; Horley Green, near Halifax, in Yorkshire; a spring at Brighton in Sussex; and Dorton in Buckinghamshire: all of these are sulphate of iron waters. Amongst the iron springs nearer London some of the following were formerly well known: Dulwich Spa; Hampstead Wells; Shadwell, near the Tower of London; Sadler's Wells, or the 'New Tunbridge Wells,' at Islington; Hoxton, Coldbath Wells, and Bermondsey Spa. Readers of Dr. Macpherson's book will find some interesting information about these once more or less popular springs of England.

In Scotland the chalybeate springs of Vicar's Bridge, near Dollar, one in Moffat, and the Hartfell Spa, near Moffat, may be mentioned as examples of sulphate of iron springs. Trefriw in North Wales (Vale of Conway Spa), 2½ miles from Llanrwst, has waters which contain much sulphate of iron and sulphate of aluminium.

Cudowa, in Prussia (Province of Silesia), lies at an elevation of 1,270 feet, near the Bohemian border. Its four alkaline chalybeate springs are all rich in carbonic acid gas. The Eugen-Quelle is richest in iron, and contains ·07 per mille bicarbonate of iron, 1·29 per mille bicarbonate of sodium, and ·0025 arseniate of iron. The good air plays a great part in the results obtained in cases of anæmia, debility, and convalescence. Ferruginous 'moor baths' and gas baths are to be had. The nearest railway station is Nachod, four miles distant, on the Breslau and Prague line.

Accommodation: Fair.

Doctors: Jacob, Scholz, etc.

Reinerz, Prussian Silesia (altitude 1,860 feet), lies in the county of Glatz, a district rich in mineral springs. Its climate is fresh, and two of its alkaline earthy chalybeate springs contain ·05 per mille bicarbonate of iron. There are walks in the neighbourhood which can be used for a 'Terrain-Cur,' after Oertel's plan. Railway stations: Rückers-Reinerz, Nachod (12½ miles), Glatz (17 miles).

Flinsberg (Prussian Silesia) lies in the Queis-Thal, on the northern slope of the Tafelfichte, at an altitude of 1,700 feet. It possesses gaseous chalybeate springs, of which the two used for drinking contain about ·04 per mille bicarbonate of iron. In the neighbourhood are walks suitable to a 'Terrain-Cur,' after Oertel's plan. The place lies in the midst of pine forest, and has a stimulant and refreshing climate. The railway station is Friedeberg (an hour's drive).

Godesberg, in Rhenish Prussia, is a favourite summer resort on the Rhine, four miles above (to the south of) Bonn. It possesses two gaseous chalybeate springs, of which the 'old' one contains ·029 per mille bicarbonate of iron, with 1·4 bicarbonate of sodium and about 1·0 common salt, whilst the 'new' spring, only used for bathing, contains more iron (·05 per mille of the bicarbonate) and less other solid constituents.

Accommodation: Good.

Doctors: Finklenburg, Oberdoerfer, Pohl, Schwann.

Driburg in Prussia (Province of Westphalia).—This spa lies at an elevation of 730 feet in a pleasant valley of the Teutoburg Forest. Of its earthy chalybeate springs the 'Hauptquelle' is the strongest, and contains ·07 per mille bicarbonate of iron, 1·4 per mille bicarbonate of calcium, 1 per mille sulphate of calcium, and much free carbonic acid gas. The 'Caspar-Heinrich-Quelle' is more weakly mineralised, and may be compared to the 'Georg-Victor-Quelle' in Wildungen. The chalybeate baths during the process of warming lose a considerable portion of carbonic acid gas. For the preparation of sulphurous peat baths the neighbouring sulphur spring of Saatz is made use of. There are two good bath establishments, the old one and the 'Kaiser Wilhelm Bad.' The station of Driburg is on the railway between Holzminden and Altenbeken, five miles from Altenbeken. The season is May 15 to October 1.

Accommodation: Fair.

Doctors: Scholz, Herrmann, Karfunkel.

Freienwalde on the Oder (Prussia), in Mark Brandenburg, is a summer resort of the people of Berlin, and contains chalybeate waters poor in carbonic acid gas.

Neustadt-Eberswalde or **Eberswalde** (Prussia), situated in a beautiful region of Mark Brandenburg (altitude 100 feet), is a summer resort, and contains chalybeate waters poor in carbonic acid gas.

Bibra (altitude 410 feet), a small summer health resort in Prussian Saxony, possesses a weakly mineralised earthy chalybeate spring, the 'Eisenquelle,' containing ·02 per mille bicarbonate of iron.

Pyrmont in Germany (Principality of Waldeck-Pyrmont).—This spa (altitude about 420 feet) lies in the beautiful valley of the Emmer, with wooded country around. It possesses cold chalybeate and muriated springs.

Of the chalybeate springs the two chief ones used for drinking (Hauptquelle and Helenen-Quelle) contain about ·07 and ·03 per mille bicarbonate of iron, 1 per mille bicarbonate of calcium, ·8 sulphate of calcium, and ·45 sulphate of magnesium; they are both rich in free carbonic acid gas, but the Brodel-Brunnen used for baths is still richer, containing about 1,540 volumes per mille of carbonic acid gas.

The muriated waters of Pyrmont contain from 7 ('Trinkquelle') to 32 ('Bohrlochsoole') per mille common salt.

The bath arrangements are good. Baths are given

both of the muriated waters and of the gaseous iron waters. The ferruginous peat of Pyrmont is made use of for 'moor baths.'

By means of the two classes of waters patients can be treated at Pyrmont for anæmia, debility, scrofula, functional nervous affections, etc. The season is from the beginning of May to October 1.

Access: By Cologne, Elberfeld, Soest, and Altenbeken; with about 18 hours of travelling.

Accommodation: Good.

Doctors: Seebohm, Gruner, Weitz, Schücking, etc.

Berka on the Ilm (altitude 770 feet), a climatic health resort in the Grand-Duchy of Weimar, possesses weak chalybeate waters. There are establishments with arrangements for pine baths, moor baths, hot sand baths, etc.

Imnau (Germany), in the principality of Hohenzollern, is pleasantly situated in the valley of the Eyach, at an elevation of 1,140 feet above the sea. Of its cold gaseous earthy chalybeate springs the richest is the Kasper-Quelle, which contains ·05 per mille bicarbonate of iron, ·03 per mille bicarbonate of manganesium, and 1·4 per mille bicarbonate of calcium. The Fürsten-Quelle, likewise rich in carbonic acid, contains only ·005 per mille bicarbonate of iron. Imnau is reached in half an hour from the railway station of Eyach.

Accommodation: Satisfactory.

Doctor: Scheef.

Liebenstein, in the Duchy of Saxe-Meiningen, lies at an altitude of 1,450 feet, sheltered on the north and north-east by the Thüringer Wald. There are beautiful walks in the surrounding forests. It is much visited by north-Germans, and contains two cold gaseous chalybeate springs and a hydro-therapeutic establishment.

The 'Altequelle' is the strongest, and, with a total mineralisation of 1·4 per mille, contains ·104 per mille bicarbonate of iron, whilst the 'Neuequelle,' with a slightly larger total of solids, contains rather less iron (·08 per mille of the bicarbonate). Season May to September.

Accommodation: Good.

Doctors: Schlaeger, Hesse.

Rippoldsau (Rippold's-Au), Grand-Duchy of Baden. Rippoldsau (altitude 1,856 feet, the best known of the 'Kniebis spas') lies in a narrow part of the Wolfthal at the south base of the Kniebis mountain. The scenery is typical of a thickly wooded Black Forest valley.

Three springs are used for drinking: the 'Wenzels-Quelle,' the 'Josephs-Quelle,' and the 'Leopolds-Quelle.' Their waters contain ·05 to ·12 per mille bicarbonate of iron, and about 1 per mille sulphate of sodium; they are cold and rich in carbonic acid gas. From the Josephs-Quelle and the Leopolds-Quelle gaseous sulphated alkaline mineral waters have been artificially prepared by the addition of sodium carbonate and carbonic acid gas; they are respectively called 'Natroine' (2·3 per mille bicarbonate of sodium and 2·4 of sulphate of sodium), and 'Schwefelnatroine' (2·2 per mille bicarbonate of sodium, 1·7 sulphate of sodium, and a little sulphuretted hydrogen gas), and are said to resemble respectively the Kreuz-Brunnen water of Marienbad, and the Schwefel-brunnen water of Weilbach.

The chalybeate waters are taken internally in anæmia and the complications of anæmia. The 'natroine' is intended for use to counteract any tendency to constipation.

For the 'iron-baths,' two springs are used somewhat poorer in iron, but richer in carbonic acid gas, than the springs used for drinking; the baths are heated by the Schwarz method.

Various hydro-therapeutic means can be employed, and there are moor-baths which are made with peat from Franzensbad in Bohemia. The season is from May 15 to September 30.

Access: By train *viâ* Strassburg and Offenburg, or Cologne and Offenburg to Wolfach, and thence 2¾ hours' drive.

Accommodation: Good.

Doctor: Martin Siegfried.

Antogast, Germany (Baden), the oldest of the 'Kniebis spas,' is situated in the Black Forest at an elevation of 1,640 feet, half an hour's drive from the railway station of Oppenau. It possesses three alkaline earthy gaseous chalybeate springs (the 'Antoniusquelle' contains ·039 per mille bicarbonate of iron).

The alkaline bicarbonates of the water are useful in atonic digestive complaints, and the mountain forest air favours the strengthening action of the iron. The springs have an old popular reputation in the neighbourhood.

Freiersbach in the Black Forest (Grand-Duchy of Baden), likewise one of the 'Kniebis group' of spas, lies in the Renchthal at an elevation of 1,260 feet. Of its cold gaseous chalybeate springs the 'Friedrichsquelle' contains ·058 per mille bicarbonate of iron, and ·013 per mille chloride of lithium, the 'Lithionquelle' contains less iron but more chloride of lithium (·017 per mille), and the 'Schwefelquelle,' smelling of sulphuretted hydrogen, is richest in iron (·1 per mille of the bicarbonate). The railway station of Oppenau is 4½ miles distant.

Griessbach, or **Griesbach** (Grand-Duchy of Baden), lies in the Black Forest at an elevation of 1,850 feet.

This, too, is one of the group of Renchthal or Kniebis spas, and possesses cold gaseous chalybeate waters, of which the 'Antoniusquelle' used for drinking is the strongest, and contains ·07 per mille bicarbonate of iron, 1·6 bicarbonate of calcium, and ·7 sulphate of sodium. The railway station of Oppenau is 7½ miles off.

Accommodation: Satisfactory.

Doctor: Frech.

Petershal (Grand-Duchy of Baden), in the Black Forest, lies at an elevation of 1,330 feet in the Renchthal on the western slope of the Kniebis mountain, five miles from the railway station of Oppenau. Its different chalybeate springs contain about ·045 per mille bicarbonate of iron, 1·5 per mille bicarbonate of calcium, and ·7 per mille sulphate of sodium.

Teinach, in a valley of the Würtemberg Black Forest, lies at an elevation of 1,310 feet under the Zavelstein. It possesses weak gaseous chalybeate springs, and a chalybeate spring poor in gas; also gaseous weakly mineralised alkaline waters which can be used as ordinary table waters. There is a hydro-therapeutic establishment.

Alexandersbad (Bavaria), on the south-eastern slope of the Fichtelgebirge, two miles from the railway station of Wunsiedel, possesses a cold gaseous alkaline earthy chalybeate spring with about ·06 per mille bicarbonate of iron; it is used for drinking and bathing. There is a hydro-therapeutic establishment, where moor and pine baths, etc., can be given. The spa is situated on the south-eastern mountain slope at an elevation of about 1,840 feet above the sea. It can be used as a climatic health resort, or 'after-cure' for patients returning from Karlsbad and Marienbad. The season lasts from May 15 to October.

Accommodation: Satisfactory.

Doctor: Müller.

Brückenau, in Bavaria (altitude 980 feet), is beautifully situated on the south-west of the Rhöngebirge, amidst forests of beech and oak. It is a four hours' drive from Kissingen.

The 'Stahlquelle' is a pleasant-tasting, weak, cold chalybeate spring (·011 per mille carbonate of iron), rich in carbonic acid gas. Besides this, there are two weakly mineralised gaseous alkaline springs, the Sinnbergerquelle and the Wernarzerquelle, used in bronchial affections, etc. The chalybeate spring is chiefly used in anæmic and debilitated women. Moor-baths and douches are employed. The season lasts from May 15 to September 30. The chief contingent of patients are women.

Accommodation: Satisfactory.

Doctors: Imhof, Wehner.

Bocklet, in Bavaria, about 4½ miles' drive from Kissingen, in a wooded and protected situation (altitude 690 feet), possesses a compound iron spring, the 'Stahlquelle,' the waters of which contain bicarbonate of iron (·088 per mille), common salt (1 per mille), and much free carbonic acid gas (temperature 50° F.). There is also a less used chalybeate spring smelling of sulphuretted hydrogen gas. The spa is of use for various anæmic and debilitated patients; also sometimes for an 'after-cure,' following treatment at Kissingen. Moor-baths are given of similar material to those at Kissingen. The season is from May 15 to the end of September.

Kohlgrub in the Bavarian Mountains, 1¼ hour's drive from the railway station of Murnau, combines the advantages of an elevated position (2,950 feet above sea level) with that of a strong chalybeate spring (the

Schmelzhaus-Quelle, used for drinking, contains ·09 per mille bicarbonate of iron). Ferruginous moor-baths are likewise employed.

Augustusbad in the kingdom of Saxony lies at an elevation of about 720 feet amidst pine woods, half an hour from the railway station of Radeberg. It possesses chalybeate waters (·02 to ·03 per mille bicarbonate of iron) and a hydro-therapeutic establishment. Ferruginous moor-baths are employed. The spa is reached from Dresden in less than an hour.

Elster (Germany, kingdom of Saxony) possesses compound iron waters, in which the action of the iron is modified by sulphate of sodium, etc. Elster has already been described in the alkaline sulphated group. (See p. 159.)

Schandau (Germany, kingdom of Saxony) is pleasantly situated (altitude 400 feet) on the Elbe in the country called the 'Saxon Switzerland.' There is sufficient accommodation, and the place is used as a summer resort. Its weak chalybeate spring contains ·015 per mille bicarbonate of iron and ·24 per mille bicarbonate of calcium. (Railway station of Schandau.)

Berggiesshuebel, a little town in the kingdom of Saxony, possesses weak chalybeate springs used for bathing by inhabitants from the neighbourhood.

Rabbi (altitude 4,100 feet) in the Austrian Tyrol lies in the Val di Rabbi, a branch of the Val di Noce. It possesses two strong alkaline iron springs, the stronger of which (the new one) is said to contain about ·18 per mille bicarbonate of iron and 1 per mille bicarbonate of sodium. The nearest railway station is San Michele (10 hours distant). The season is from the middle of June to the middle of September.

Pejo in the Austrian Tyrol lies in the Pejo valley to

the south of the Ortler district at 4,430 feet above the sea-level. It possesses an alkaline chalybeate spring containing, according to Bizio, ·05 per mille bicarbonate of iron. The nearest railway station is San Michele (12 hours' drive).

Andelsbuch (Austrian Tyrol) a summer resort in the Vorarlberg, lies at an altitude of 1,970 feet, and possesses a gaseous chalybeate spring.

Antholz (Austria) a summer resort situated at an altitude of 3,600 feet in a southern Tyrolese valley, possesses a chalybeate spring and a sulphur spring, both of which are used for bathing by inhabitants of the neighbourhood.

Franzensbad (Bohemia).—This spa, some of whose springs, especially the Neuquelle and the Stahlquelle, yield important compound chalybeate waters, has been described in the chapter on sulphated alkaline waters (see Chapter XI., p. 154), to which group most of its springs belong.

Marienbad in Bohemia possesses chalybeate waters of moderate strength, but, for convenience, has been described in Chapter XI. amongst spas with sulphated alkaline waters. (See p. 151.)

Liebwerda (altitude 1,420 feet), in the north of Bohemia, lies on the south-western slope of the Tafelfichte, half an hour's distance from the railway of Raspenau-Liebwerda. It possesses the 'Stahlbrunnen,' which contains ·03 per mille bicarbonate of iron and small quantities of alkaline earthy salts. The 'Christians-brunnen' is a weakly mineralised, alkaline-earthy, gaseous spring, whose water can be used with meals or as a simple refreshing draught. Moor-baths are likewise employed.

Koenigswart, a health resort in Bohemia (altitude

2,230 feet), is a railway station on the line from Eger to Pilsen, five miles before the station of Marienbad. The Curhaus lies on a wooded hill half an hour's drive from the station. The cold gaseous chalybeate springs are said to contain ·08 per mille bicarbonate of iron. The Ricardsquelle is a simple gaseous spring. Moor-baths are made use of. The position on a mountain slope looking towards the south and the purity of the air are favourable conditions for the treatment of anæmia and convalescence, and for an after-cure to Marienbad, etc.

Bartfeld (altitude 1,000 feet) in Hungary lies in a pleasant valley at the foot of the Kamenahola, a spur of the Carpathians. It possesses several cold gaseous alkaline muriated chalybeate springs, containing a small amount of iodide of sodium. The 'Doctorquelle' has 4·8 per mille bicarbonate of sodium, 1·1 per mille common salt, ·05 per mille bicarbonate of iron, and ·001 per mille iodide of sodium. There are arrangements for hydro-therapeutic treatment. Anæmic conditions, especially those associated with scrofula or with dyspeptic symptoms, are treated here. The spa may be used as an after-cure to Karlsbad, Marienbad, etc. It is five hours distant from the nearest railway station, Eperies, and half an hour from the town of Bartfeld.

Elöpatak or **Arapatak** (Transylvania) possesses cold strong alkaline chalybeate waters, rich in free carbonic acid gas. It is the most frequented spa in Transylvania, and lies in a pleasant sheltered valley, 12½ miles distant from Kronstadt, at an elevation of about 2,030 feet above the sea level. Of the various springs the two most used internally are the Stammbrunnen, containing about ·17 per mille carbonate of iron, and the Neubrunnen, containing about ·24 per mille.

The waters are taken in chlorosis and in menstrual and digestive disorders of anæmic persons. Hydro-therapeutic processes can be likewise employed. The season lasts from the middle of May to the end of September.

Borszek, a Transylvanian health resort, is situated at an altitude of about 2,890 feet in the Carpathian Mountains, near the Roumanian frontier, and possesses cold, alkaline-earthy, chalybeate springs, of which the Kossuthquelle is richest in iron. Moor-baths are made use of.

Acquarossa (Switzerland, Canton Ticino), at an altitude of 1,150 feet, is beautifully situated amongst high mountains in the Val Blenio about 1½ hour's drive from the station of Biasca on the Italian side of the St. Gothard railway. The waters have a temperature of 77° F., and according to Koerner's analysis contain ·034 per mille bicarbonate of iron, ·019 bicarbonate of manganesium, ·00024 arseniate of calcium, ·0025 borate of magnesium, ·0046 chloride of lithium, and 1·1 sulphate of calcium. (The total mineralisation is 2·5 per mille.) The waters deposit a red ferruginous, muddy material, from which they have derived their name, and which is heated and applied externally in the treatment of chronic cutaneous affections.

Tarasp (Switzerland, Grisons) possesses compound chalybeate springs, of which the 'Bonifaciusquelle' is the strongest, containing ·045 per mille bicarbonate of iron, together with bicarbonates of sodium and calcium. See in Chapter X. (p. 157), where Tarasp is described amongst the sulphated alkaline group.

Tiefenkasten and **Solis,** near **Alveneu** (Switzerland, Grisons), possess sulphated alkaline chalybeate springs. These are referred to under ALVENEU amongst sulphur. (See p. 234.)

Andeer-Pignieu.—Andeer in Switzerland (Grisons) lies in the Schamserthal at an altitude of 3,200 feet, about five hours' drive from the railway station of Chur. The water of the neighbouring spring at Pignieu is conducted to Andeer; it is a weakly mineralised earthy (1·7 per mille sulphate of calcium) water, containing a small amount of bicarbonate of iron (temperature 66°–68°F). Ferruginous mud-baths are likewise made use of. The season lasts from the middle of June to the end of September.

San Bernardino (Switzerland, Grisons) is situated at an altitude of 5,320 feet, on the road from Splügen to Bellinzona, about eleven hours by diligence from the railway station of Chur and seven and a half from Bellinzona. Its cold gaseous earthy chalybeate spring, according to De Planta's analysis, contains ·035 per mille bicarbonate of iron, ·01 bicarbonate of strontium, and 1·2 sulphate of calcium, its total of solid constituents being 2·59 per mille. Accommodation is just fair.

Fideris (Switzerland, Canton Grisons) lies at an altltude of 3,460 feet in the Praettigau Valley, one hour from the station of Fideris, on the railway between Landquart and Davos. It possesses gaseous weak chalybeate waters (·01 per mille bicarbonate of iron, with a total mineralisation of 1·9 per mille), resembling the class of 'table waters.' The climate plays a chief part in the treatment.

Other chalybeate springs in Switzerland are those of Passugg (see p. 136); Farnbühl (altitude 2,310 feet) in Canton Lucern, an hour from the railway station of Malters; Gonten (altitude 2,900 feet) in Canton Appenzell; Rothenbrunnen in Canton Grisons, with a weakly mineralised compound chalybeate water, and an elevation of 2,000 feet above sea level; and, lastly, Morgins in Canton Valais, three and a half hours' drive from the railway station of Monthey, with the high elevation of 4,300 feet, but with 2·4 per mille sulphate of calcium in its waters.

Lamalou (France, Department of Hérault) lies at an altitude of 620 feet in a deep valley of the Cevennes Mountains. It is a station on the railway from Bédarieux to Castres. The springs are situated in three groups: Lamalou-le-Bas, Lamalou-le-Centre, Lamalou-le-Haut; they have temperatures ranging from 59° to 117° F., and may be classed as chalybeate and weak alkaline.

The 'Source Capus' of Lamalou-le-Centre (temperature 59° F.) contains ·05 per mille bicarbonate of iron and ·001 per mille arseniate of sodium. The so-called 'Source Petit-Vichy' (temperature 61·5° F.) of Lamalou-le-Haut has a total of only 1 per mille solids (carbonate of sodium, etc.), and is very rich in carbonic acid gas; it may, therefore, be considered as a simple aërated water similar to 'table waters.' The Source de la Vernière' contains 1·1 per mille bicarbonate of sodium, but none of the springs are strongly alkaline.

A chamber cut in the rock where the 'Source Usclade' of Lamalou-le-Bas (temperature 117° F.) arises, has been made use of as a vapour bath (temperature 122° F.).

Lamalou is resorted to for chronic rheumatism, neuralgias, incipient tabes dorsalis, and chronic affections of the nervous system. The season is from May 15 to October 15.

Doctors: Belugou, Cros, etc.

Audinac, a French village (Department of Ariège), in the Pyrenees, one and a quarter mile from the railway station of St. Lizier, possesses weakly mineralised earthy chalybeate waters (temperature 71·6°–72·8° F.).

Bagnères-de-Bigorre (France, Hautes-Pyrénées) possesses some chalybeate springs. This spa is described in the earthy group (see p. 256).

Rennes-les-Bains (France, Department of Aude).—The village is situated in a narrow valley at an altitude of 1,040 feet on the banks of the River Salz. It possesses thermal chalybeate springs, the hottest is the 'Source du Bain Fort' (temperature 124° F.), which, out of a total of 1 per mille solid constituents, contains about ·031 per mille carbonate of iron. The 'Source de l'Eau du Cercle' (temperature 54° F.) contains ·015 per mille sulphate of iron.

There are various weak muriated springs at Rennes, and the River Salz, as its name seems to imply, is itself a mineral water, containing a total of about 5 per mille solids (chlorides of sodium and magnesium, sulphates of calcium, sodium and magnesium). Rennes is resorted to by feeble and anæmic rheumatic patients, etc.

Barbotan (France, Department of Gers).—Is a quiet village near Cazaubon, 19 miles from the railway station of Mézin, and possesses thermal chalybeate springs (temperature 59° to 100° F.), containing ·03 per mille carbonate of iron and a minute amount of sulphuretted hydrogen. The waters are chiefly employed for making mud baths, used in cases of chronic rheumatism and joint affections. The spa is visited mostly by patients from the neighbouring part of France. The season is from the commencement of June to the end of September.

Amphion-les-Bains (France, Savoie) lies on the southern shore of the Lake of Geneva (altitude 1,240 feet), and is connected by a tramway with the railway station of Evian-les-Bains, about two miles distant. It possesses a weakly mineralised water, rich in carbonic acid gas, and containing the minute quantity of ·006 per mille phosphate of iron, with very small quantities of the bicarbonates of calcium, magnesium, and sodium.

La Bauche (France, Department of Savoie).—Half an hour's drive from the railway station of Lépin-Aiguebelette, lies at an elevation of 1,970 feet in a fertile valley of Savoy on the slope of the Signal Mountain. Its cold, chalybeate water is said to contain ·14 per mille bicarbonate of iron, and ·03 per mille crenate of iron.

Charbonnières (France, Rhone).—A village six miles north-west of Lyons, possesses cold chalybeate waters (·04 per mille bicarbonate of iron), poor in free carbonic acid gas.

Forges-les-Eaux (France, Department of Seine-Inférieure).—The town (altitude 525 feet) lies on the railway from Paris to Dieppe, *viâ* Pontoise, and owed its reputation to the visit in 1632 of Louis XIII. with his wife Anne of Austria, and his famous minister Cardinal Richelieu. The waters of the 'Source Cardinale' contain ·098 per mille crenate of iron, with small quantities of alum and earthy salts.

Orezza lies at an altitude of 1,960 feet amongst the mountains of the north-eastern part of the island of Corsica. It possesses two strong gaseous chalybeate springs, containing ·128 per mille bicarbonate of iron, with a total of ·85 per mille solids.

Santa Catarina (Upper Italy).—About three miles from Bormio, at an altitude of about 5,600 feet, has strong chalybeate waters, with a climate analogous to that of St. Moritz in the Upper Engadine.

Recoaro (Italy, Province of Vicenza) lies at an altitude of 1,400 feet, and is about 26 miles from the railway station of Vicenza, with which it is connected by a steam-tramway. Of its many chalybeate springs the best known is the 'Lelia,' which, according to Bizio, contains ·046 per mille carbonate of iron, together with small amounts of carbonate of calcium and of the sulphates of calcium and magnesium; it is rich in carbonic acid gas. Accommodation is good.

We shall now consider foreign waters containing sulphate of iron, the English ones having been already mentioned amongst English chalybeate waters.

Alexisbad (Germany, Duchy of Anhalt) lies in the Selkethal at the foot of the lower Harz Mountains, two hours from the railway station of Gernrode. The chalybeate waters used for drinking are the Alexis-Brunnen and the Freundschafts-Brunnen, containing bicarbonate and sulphate of iron. The Selke-Brunnen, containing chloride (·1 per mille) and sulphate (·05 per mille) of iron and sulphates of sodium, magnesium, and calcium, is used for bathing. 'Soolbäder' and pine baths, douches and massage can also be obtained. The spa has a pleasant sequestered position, at an altitude of about 1,080 feet above the sea-level. The air is fresh and rather moist, and there are pleasant shady walks in the neighbouring woods. The season lasts from the beginning of June to September 15.

Hermannsbad at **Muskau** in Prussian Silesia.—Muskau (altitude 320 feet) on the Neisse, in the Oberlausitz, a station on the branch railway from Weisswasser, possesses cold sulphate of iron waters. The 'Trinkquelle' is said to contain about ·19 per mille sulphate of iron, ·24 per mille bicarbonate of iron, and ·5 per mille sulphate of calcium, whilst the stronger 'Badequelle' has ·75 sulphate of iron, ·54 bicarbonate of iron, and 2·08 sulphate of calcium. Ferruginous moor-baths are to be obtained. Hermannsbad lies in the middle of the celebrated park and gardens of Prince Pückler.

Hermannsbad near **Lausigk**, in the kingdom of Saxony, possesses strong sulphate of iron waters (over 4 per mille), unsuitable for internal use.

Ratzes (in the Austrian Tyrol, altitude 3,900 feet) lies in a wooded ravine close to the Schlern Mountain. It possesses a sulphate of iron spring (·3 per mille sulphate of iron), and a cold sulphur spring. The nearest railway station, Atzwang, is three and a quarter hours distant.

Mitterbad (Austrian Tyrol, altitude 3,110 feet), three and a half hours' distance from Meran, lies in the romantic Marau Valley, and possesses a chalybeate spring, containing sulphate of iron, with minute quantities of arsenic and of the sulphates of manganesium, strontium, zinc, and copper.

Parad (Hungary, altitude 660 feet), a station on the railway from Kis-Terenne to Kaal-Kapolna, possesses sulphate of iron waters, the strongest of which is said to contain 5·5 per mille sulphate of iron, and 3·03 per mille sulphate of aluminium. In the neighbourhood is the Cseviczequelle, a sulphurous spring containing 1·1 per mille carbonate of sodium, much carbonic acid, and ten volumes per mille sulphuretted hydrogen gas. Further off (two hours' distance) is the Clarissequelle, containing ·06 per mille bicarbonate of iron.

The sulphate of iron springs of Levico, Roncegno, Vals, and Erdöbenye, will be described amongst the arsenical waters. Amongst other sulphate of iron springs there are those of Ronneby, the best known spa in Sweden; the 'new spring,' containing about 2·5 per mille sulphate of iron and 1·5 sulphate of aluminium, is only used for bathing, whilst the weaker 'old spring' (·33 per mille sulphate of iron, and ·38 sulphate of aluminium) is sometimes used internally.

RIO (Elba) has a sulphate of iron spring. Another interesting one is the hot PISCIARELLI spring near Pozzuoli, containing both sulphate of iron and alum, described by Pliny and still used, it is said, by Neapolitans as an external application. There are many other sulphate of iron waters in Italy.

AUTEUIL, in France, a suburb of Paris, possesses a cold chalybeate spring, containing ·71 per mille of sulphate of iron and aluminium, and 2 per mille of the sulphates of calcium, magnesium, and sodium, with traces of arsenic. The total mineralisation is 3·2 per mille. PASSY, a part of Paris, also contains sulphate of iron springs, though not employed at present; the analysis of two of them shows the presence respectively of ·045 and ·41 per mille sulphate of iron.

The chalybeate (carbonate of iron) waters of VAL SINESTRA, CERESOLE REALE, VIC-SUR-CÈRE, SYLVANES and BUSSANG, are described in the arsenical group. Amongst other chalybeate waters may be mentioned those of:— HAARLEM, in Holland; HITZACKER-WEINBERG, in Hanover; MALMEDY, in Rhenish Prussia, near Belgium; the DINKHOLDER-BRUNNEN, near Braubach, on the Rhine; ALBERSDORF, on the Baltic Canal (ferruginous baths and hydro-therapeutic establishments); HOFGEISMAR, in the Prussian Province of Hesse-Nassau (rather weak); STETTIN and POLZIN in Pomerania (poor in CO_2); RONNEBURG, in the Duchy of Saxe-Altenburg (rather poor in CO_2); LOBENSTEIN (poor in CO_2), in the principality of Reuss-Schleiz; NIEDERNAU, in the Würtemberg Black Forest; RASTENBERG, in Thuringia (Saxe-

Weimar); REIBOLDSGRÜN, better known for its sanatorium for consumptive patients (altitude 2,250 feet), and LINDA, in the Kingdom of Saxony; KÖNIG OTTO BAD (near WIESAU), KELLBERG, and STEBEN, in Bavaria; LANGENAU or NIEDERLANGENAU, CHARLOTTENBRUNN, BUKOWINE, ALT-HEIDE, HERMSDORF, near Goldberg, LAUCHSTÄDT (poor in CO_2), and SCHWARZBACH (weak), in Prussian Silesia; KARLSBRUNN (altitude 2,520 feet) in Austrian Silesia; STERNBERG and SANGERBERG in Bohemia; PYRAWARTH (·11 per mille bicarbonate of iron, but comparatively weak in CO_2) in Lower Austria; VELLACH, or FELLACH (altitude 2,750 feet), in Carinthia; TARCSA, or TATZMANNSDORF (with sulphate of sodium), in Hungary, near the borders of Lower Austria and Styria; SZLIACS, and other springs near ALTSOHL (some of them are thermal), in Hungary; KORYTNICA and BOESING, in the Carpathians in Hungary; KRYNICA, in the Carpathians in Galicia (altitude about 2,000 feet, gaseous earthy fairly strong chalybeate waters, moor-baths, etc.); and ZAIZON in Transylvania (see Chapter XVI. p. 266). ALTWASSER, in Prussian Silesia, was well known as a chalybeate spa up to 1869, in which year the springs were greatly damaged by the coal-mining work.

CHAPTER XIII

ARSENICAL WATERS

Arsenic occurs in appreciable quantities in some mineral waters, so that from their employment one may expect an alterative action in cases of anæmia and various cachectic conditions, especially malarial cachexia, where ordinary chalybeate waters do not act well. These waters may likewise be of use in some chronic skin affections.

The action of arsenic is associated with that of sulphate of iron in the waters of Roncegno and Levico in Italy; Erdöbenye, in Hungary; and the sulphate of iron waters of Vals, in France; with bicarbonate of iron in the waters of Ceresole Reale, Val Sinestra, Vic-sur-Cère, Sylvanes, and Bussang; and with that of muriated alkaline waters at La Bourboule. It may be doubted whether the amount of arsenic in the weaker members of this group can exert any special therapeutic action. The weakly mineralised waters of Mont Dore contain the minute amount of ·0009 per mille arseniate of sodium: that is, under one milligramme in the litre.

Besides the waters separately mentioned in this chapter, minute quantities of arseniate of sodium occur in the waters of Royat (·0045 per mille in the 'Source Saint-Victor'), Vichy (·002 in the 'Source Grande Grille'), Uriage (·002), Lamalou (·001 in the 'Source Capus'), and traces of arsenic are found in the waters of Bath, Baden-Baden, Kreuznach, Plombières, St. Honoré, Poretta (in Italy), etc.

Court-Saint-Etienne (Belgium, Brabant) possesses a well, discovered in 1878, said to contain as much as ·0097 per mille arsenious acid, or ·0263 per mille of arseniate of sodium, out of a total of only ·28 per mille solids. The water is used for exportation solely.

La Bourboule (France, Department of Puy-de-Dôme).—La Bourboule is situated in a pleasant Auvergne valley, at an elevation of 2,780 feet above the sea, on both banks of the River Dordogne.

The waters of La Bourboule are distinguished from other muriated alkaline waters by the amount of arsenic they contain. The two principal springs, the Source Perrière and the Source Choussy, have a temperature of 130° F. (140° F. before being pumped up), and the first of them contains 2·8 per mille of both the chloride and the bicarbonate of sodium and an amount of arsenic equivalent to ·028 per mille of the arseniate of sodium. The two Sources Fenestre are cold (66° F.) and much more weakly mineralised; they are used for baths to lower the temperature of the other springs.

The waters are employed for drinking, for baths, and for inhalation. They may be used in affections of the respiratory system, and in cases where muriated alkaline waters are indicated, but owing to the amount of arsenic which they are said to contain a good result may also be expected in many scrofulous and cachectic conditions, chronic malarious troubles, feeble rheumatic and gouty patients, and chronic skin affections, when arsenic is indicated. They have been found useful, it is said, in some cases of glycosuria and albuminuria, and have also a certain reputation in early stages and quiescent conditions of pulmonary tuberculosis.

The season is from May 25 to September 30.

Access: By Paris, Orleans, and Montluçon to Laqueuille; omnibus in one and a half hour from the railway station of Laqueuille.

Accommodation: Good.

Doctors: Vérité, Gilchrist, Noir, Morin, etc.

Mont Dore (France, Department of Puy-de-Dôme).—Mont Dore lies on the Dordogne, in a picturesque part of the Auvergne Mountains, at an elevation of 3,440 feet above the sea-level. The thermal springs, which, judging from remains found in the neighbourhood, were evidently known to the Romans, yield weakly mineralised waters, having a temperature of 108°–113° F., and containing a minute but appreciable amount of arseniate of sodium (about ·001 per mille). The thermal springs differ chiefly in the amount of carbonic acid gas they contain. One of the springs is cold, the 'Fontaine Sainte-Marguerite' (temperature 54° F.), and is rich in carbonic acid gas.

The waters are used for baths, drinking, and inhalation. The spa guests include many professional people who have to speak in public, or sing, and are suffering from chronic laryngitis, bronchitis, or other chronic affections of the respiratory organs. In their case the inhalation treatment is much employed. Many of the patients are gouty or rheumatic, and some of them come for functional nervous troubles.

Mont Dore enjoys a special reputation in the treatment of asthma, and it may be said that more cases are benefited there than at any other place. The success is most frequent in asthma complicated with chronic bronchial catarrh, but not rarely true nervous asthma likewise derives benefit. Relapses, however, are not uncommon.

Much of the good result is due to the climate and ordinary thermal treatment. The arsenic is supposed likewise to have some influence. In the bath arrangements there is still room for improvement, which is likely to be effected at an early period. The season is June 15 to September 15.

Access: Six hours' drive from the railway station of Clermond Ferrand; or better, direct from Paris by the Orleans railway to Laqueuille, and thence one and a half hour's drive. Next year the railway from Laqueuille to Mont Dore will be ready for use.

Accommodation: Good.

Doctors: Emond, Cazalis, Mascarel, Schlemmer, Joal, Léon Chabory, Nicolas, etc.

Levico, a village in the Austrian Tyrol, lies at the entrance of the beautiful Sugana valley, at an elevation of 1,700 feet above the sea, twelve and a half miles to the east of Trento. It possesses cold sulphate of iron and arsenic waters, used internally for anæmic conditions, malarial cachexia, etc., and externally for catarrhal conditions of the female generative organs, etc. It is in the grottoes of Vetriolo, almost 4,900 feet above the sea on the southern slope of the Monte Fronte, that the strong and weak springs of Levico are situated. Near the grottoes is the bath-establishment of Vetriolo.

The weak Levico water (·66 per mille sulphate of iron, ·00095 arsenious acid) is taken internally, at the commencement of the course, in doses of one or two tablespoonfuls twice or thrice a day, during or after meals. After two or three weeks the stronger water is used in similar doses, which may be doubled later on.

The strong Levico water contains ·009 per mille arsenious acid, 5·13 per mille sulphate of iron, ·05 per mille sulphate of copper, and ·65 per mille sulphate of aluminium.

The season asts from June 1 to the end of September.

Roncegno, a village in the Austrian Tyrol, nineteen miles east of Trento, possesses strong sulphate of iron and arsenic waters. It lies in the Val Sugana at an altitude of 1,750 feet, has accommodation for visitors, and will shortly have a railway station on the new Val Sugana line.

According to Professor Manetti's analysis the Roncegno waters contain ·067 per mille arsenious acid, 2·03 per mille ferric sulphate (persulphate of iron), ·38 per mille ferrous sulphate (protosulphate of iron), ·027 per mille sulphate of copper, 1·27 per mille aluminium sulphate, and 1·63 per mille organic matter. One to four tablespoonfuls of the water are taken daily, the maximum daily dose being six tablespoonfuls.

The season lasts from May 1 to the end of September.

Ceresole Reale (Italy, Piedmont) is a small village in the valley of the Oreo, which can be reached in about five hours from Turin. It lies at an elevation of 5,290 feet between the Grand Paradis and the Levanna Mountains, both of them over 11,000 feet. The water of the two springs is similar, and contains, according to Sobrero's analysis, ·17 per mille bicarbonate of iron, ·0057 per mille arseniate of sodium, and about ·003 per mille each of bicarbonates of lithium and manganesium. The situation is beautiful, the air bracing, and the accommodation good.

Val Sinestra (Switzerland, Canton Grisons).—The springs of Val Sinestra are about three hours distant from Tarasp-Schuls, and the waters may be obtained at Schuls, freshly brought there each day from the spring. They are said to contain about one-fifth of the amount of arsenic held in the strong waters of Levico, and to have the advantage of containing bicarbonate instead of sulphate of iron. According to Husemann's analysis the Ulrichsquelle contains ·0017 and the Conradinsquelle ·0019 per mille of arseniate of sodium. Both these springs include amongst their constituents about ·03 per mille of the bicarbonate of iron and about 1·5 per mille of the bicarbonate of calcium.

Cudowa, in Prussian Silesia (see p. 179).—Arsenic has been found in the Eugenquelle in the form of arseniate of iron. The water of this spring is said to contain ·07 per mille bicarbonate of iron, 1·29 per mille bicarbonate of sodium, and ·0025 arseniate of iron.

Erdöbenye (Hungary) lies at an altitude of about 780 feet in a well-wooded valley three miles from the railway station of Liszka-Tolcsva. Its waters contain ·34 per mille sulphate of iron, 1·8 per mille alum, and some arsenic. We have classed these waters with hesitation, as we do not know if the amount of arsenic present justifies them being placed in this group.

Bussang (France, Department of Vosges) is situated in a valley of the Vosges Mountains at an elevation of 2,200 feet above the sea. The railway station is the terminus of a branch line from Epinal. Its weak alkaline waters are rich in carbonic acid gas, and out of a total of 1·5 per mille solids, contain ·0029 per mille of carbonate of iron, ·0029 per mille of carbonate of manganesium, and ·0012 per mille arseniate of iron. The water is chiefly exported and drunk at meals.

Sylvanes (France, Department of Aveyron) lies in a mountainous district at an altitude of 1,312 feet, and is reached from the railway station of Ceilhes-Roqueredonde. Its thermal waters (88° to 97° F.) out of a total of 1 per mille solids contain ·02 carbonate of iron and ·016 of arseniates (iron and magnesium).

Vals (France, see p. 132), noted for its cold alkaline springs, possesses also weakly mineralised sulphate of iron springs. The 'Source Saint-Louis,' out of a total of ·4 per mille solids, contains ·04 per mille sulphate of iron and ·001 per mille of arseniates. The 'Source Dominique' contains ·003 per mille arseniate of iron.

Vic-sur-Cère (France, Department of Cantal), lying at an altitude about 2,200 feet at the foot of the Cantal Mountains, possesses cold, gaseous, chalybeate waters, which contain ·05 per mille bicarbonate of iron, ·008 per mille arseniate of sodium, 1·8 per mille bicarbonate of sodium, 1·2 per mille of earthy bicarbonates, and 1·2 per mille of common salt. The establishment lies about three-quarters of a mile off the village.

CHAPTER XIV

SULPHUR WATERS

Although it is difficult to explain the action of sulphur waters, and although their special therapeutic value has been altogether called in question, yet experience seems to show that the stronger waters of this group, at all events, indubitably exercise some therapeutic effect. The action of the very weak sulphur waters is probably similar to that of simple thermal or ordinary hydrotherapeutic treatment, aided by climate, diet, healthy mode of life, and general medical guidance. In compound sulphur waters the action is modified by the presence of other constituents in the waters.

Sulphur waters are used for bathing, drinking, and inhalation. The complaints treated are chronic metallic poisoning, hæmorrhoids, and bronchial, laryngeal, and pharyngeal catarrhs; also rheumatism, gout, constitutional syphilis, and chronic skin eruptions.

The action of prolonged sulphur baths in skin affections is probably similar to that of simple thermal baths or earthy thermal baths (see p. 49), but in some cases the sulphur may exercise a special germicidal action on the skin.

Sulphur baths are used at Aix-la-Chapelle, Barèges, and other localities to bring out any signs of latent syphilis (see under Aachen, p. 213).

An argument in favour of the special action of the

stronger sulphur waters when taken internally is that prolonged courses are apt to induce emaciation and anæmia. In thin feeble persons their internal use is not advisable.

Sulphur waters as a general rule keep badly, and should therefore, if possible, be drunk at the spring itself, in spite of the improved methods of bottling. In the waters rich in sodium sulphide a rapid change takes place on exposure to the air. The carbonic acid in the water and the air combines with the sodium to form carbonate, and a part of the sulphur thus set free constitutes the flocculent precipitate seen at Bagnères-de-Luchon, etc., whilst another part combines in the nascent condition with hydrogen from the water, and is disengaged in the form of sulphuretted hydrogen gas (the characteristic smell of which is sometimes not noticed when the water is quite fresh). Another part of the sulphide of sodium undergoes conversion successively into the hyposulphite, sulphite, and sulphate of sodium. In this condition the water is termed a 'degenerated sulphur' water, and is alkaline, owing to the presence of sodium carbonate. When calcium sulphide waters are exposed to the air, calcium carbonate is formed, and sulphuretted hydrogen set free.

We shall first describe Cauterets, Bagnères-de-Luchon, Eaux Bonnes, Aix-la-Chapelle, and Aix-les-Bains, and shall then proceed with the other spas of this group in geographical order.

Bagnères-de-Luchon (France, Haute-Garonne).—Luchon ('Balnearœ Lixonienses') is beautifully situated (at an altitude of 2,050 feet) in the broad valley of Luchon, near to the Spanish frontier. The climate is mild, though, owing to the mountainous position, the changes in temperature are very sudden.

The forty-nine thermal sulphur springs vary in temperature from 61° to 150° F., and the amount of sulphide of sodium also differs much: that in the Source de la Reine amounts to ·056 per mille, and that in the Source Bordeu to ·07 per mille. The waters become rapidly altered on exposure to the air, and throw down a precipitate of sulphur. They are used for drinking, but chiefly for baths. By employing the springs of different strengths the baths may be made more or less excitant, as required.

The affections treated at Luchon include chronic rheumatism and the resulting stiffness in joints, torpid scrofulous conditions, chronic cutaneous eruptions, and syphilis. In affections of the respiratory organs inhalation of the waters is employed. The bath establishment and hydro-therapeutic arrangements are excellent.

Besides the many sulphur springs there are some chalybeate springs, and an alkaline one in the neighbourhood. The season is from June 15 to October 15.

Access: Viâ Paris, Limoges, Toulouse and Montréjeau to the railway station of Luchon, or *viâ* Bordeaux and Montréjeau.

Accommodation: Good.

Doctors: Doit-Lambron, Estradère, Garrigou, Ferras, De Lavarenne, Serrand, etc.

Cauterets (France, Department of Hautes-Pyrénées).—Cauterets is a little town situated in the narrow winding valley of the Gave, at an elevation of about 3,200 feet above the sea-level.

Of the many thermal sulphur springs the six chief ones have temperatures ranging from 103° to 128° F., and out of a total of about ·2 per mille solids, have amounts of sulphide of sodium varying from ·01 to ·022 per mille. The principal spring is the 'Source la

Raillère,' on the banks of the Gave, about which crowds of spa-guests assemble during the season in the early morning. The waters (temperature 103° F.) contain ·015 per mille sulphide of sodium, and have a particular reputation in affections of the respiratory organs. The hottest spring, 'Source des Œufs,' supplies the principal bath establishment of Cauterets, called the 'Thermes des Œufs,' which is well furnished as to baths, douches, etc. The 'Source Manhourat' (temperature 121° F., containing ·01 per mille of sulphide of sodium) is only used for drinking, and has a special reputation in cases of dyspepsia and uric acid gravel. Inhalation is employed for chronic pharyngitis, laryngitis, etc.

The springs 'de César' (temperature 118·5° F.), 'des Espagnols' (temperature 116° F.), and 'Pause vieux' (temperature 107·2° F.) are likewise reckoned amongst the chief. The waters of Cauterets are said to be less excitant than those of most of the central Pyrenean spas. The season is from May 15 to November 1.

Access: By Paris, Bordeaux, Tarbes, and Lourdes to Pierrefitte; one and a half hour's drive from the railway station of Pierrefitte.

Accommodation: Good.

Doctors: Duhourcau, Flurin, Bouyer, E. Michel, Lahillone, etc.

Eaux Bonnes (also called BONNES) (France, Department of Basses-Pyrénées).—The village (altitude 2,460 feet) is situated on a rocky terrace in the picturesque valley of Ossau, 26 miles to the south of Pau.

Of the thermal sulphur springs (temperature 72°–91·5°) the hottest and most important is the 'Source Vieille' (total solids amount to only ·6 per mille; about ·02 per mille sulphide of sodium; traces of other sulphides; about ·3 per mille chloride of sodium and other chlorides; flakes of 'barégine').

The waters are used chiefly in chronic bronchitis, granular pharyngitis, and catarrhal affections of the respiratory organs. The supply of waters is not very great, and they are employed chiefly for drinking, sometimes for inhalation, and but little for baths, whereas the waters of Eaux-Chaudes (five miles off) are used principally for bathing and external applications.

The effect of the waters is excitant at first, augmenting the secretion from the mucous membranes and the amount of cough. The urine is increased in amount, the pulse becomes more frequent, and the appetite greater. With this general excitation and exaggeration of symptoms at the commencement of the treatment there may be a feeling of general uneasiness and insomnia, but all this should pass off, and an improvement in the symptoms should take place, or the normal condition should be reached.

Small doses are used at the commencement (half a glass or less), which are gradually increased, but more than three or four glasses are seldom prescribed. Whey may be mixed with the water. The treatment is contraindicated when there is fever or acute inflammation, in neurotic (dry) asthma, and in very irritable subjects. The 'Source Froide' (temperature 53·6°) is used in dyspeptic conditions.

The season lasts from June 1 to September 30. Warm clothing should be brought, owing to the fluctuations of the temperature in the Pyrenees. Patients often go to the seaside for an after-cure.

Access: 4 miles from the railway station of Laruns, the terminus of a branch line from Pau.

Accommodation: Good.

Doctors : Cazaux, Leudet, Valéry-Meunier, etc.

Aachen (AIX-LA-CHAPELLE) (Germany, Rhenish Prussia).—Aachen (altitude 530 feet) is a very important industrial city in Rhenish Prussia with 96,000 inhabitants, and its size in some respects modifies the character of the place as a spa. In its cathedral are the famous relics of Charles the Great, who has been honoured as the discoverer of the springs and founder of the town, but thermal waters at Aachen were certainly known to the Romans, and were visited in A.D. 756 by King Pepin the Short, at which time the town bore the name of Aquisgrani. It was in 1267, when Richard[1] Earl of Cornwall was King of the Romans, that the 'King's bath' came into the possession of the town.

The town is built on sandy soil, and is fairly sheltered by hills ; the climate is moderately moist, and the average temperature is higher than that of Berlin in winter and lower in summer. Its surroundings are beautiful. The Lousberg on the north is only a short walk from the town, and on the south-west the cool promenades of the Aachener Wald (the property of the town) can easily and quickly be reached by the electric tram.

The different springs lie in the middle of the town, including the Kaiserquelle (strongest ; temperature 131° F.), Quirinusquelle, Rosenquelle, Corneliusquelle, and others. The waters of the various springs are very similar in mineral constituents, containing about 2·6 to 2·8 per mille common salt, and about ·6 per mille carbonate of sodium ; they differ from each other, however, in temperature (113°–133° F.), and in the amount of sulphur (sulphide of sodium and sulphuretted hydrogen) which they contain.

[1] Richard likewise presented to the town the special regalia sent over from England for his coronation.

The Elisen-brunnen derives its water from the Kaiserquelle, and is the one most used for drinking. The water of the Kaiserquelle, etc., artificially freed of sulphur, and impregnated with carbonic acid gas, is sold in bottles as an agreeable aërated table water. At the bath establishments there are arrangements for vapour baths, and various hydro-therapeutic processes, especially the combined douche and massage. In the town there is likewise a 'Zander Institution,' furnished with Dr. Zander's medico-mechanical appliances for Swedish gymnastics.

On account of the chloride of sodium they contain, the waters are taken in catarrhal conditions of the stomach and alimentary canal, and of the bronchi; there are inhalation chambers for bronchial and laryngeal affections. Chronic rheumatism and gouty cases, and the stiffness of joints resulting from these affections, are much treated by douches, massage, etc. A combination of douche and massage is employed at Aachen, similar to the 'Aix douche' at Aix-les-Bains (*q.v.*), but the treatment at Aachen is carried out by a single attendant in most cases.

Chronic skin diseases, such as eczema and psoriasis, are treated at Aachen with some success, the results obtained being in some cases partly due to medicaments, such as chrysophanic acid, etc., applied at the same time.

It is, however, in the treatment of syphilis that Aachen has become most noted. Seventy per cent. of the patients visiting the spa are syphilitics. The baths are doubtless useful adjuncts in the treatment, but it is by careful and methodical anti-syphilitic medication that the doctors at Aachen have secured this great reputation for their spa.

The sulphur treatment is esteemed by some authorities as occasionally of service to bring out signs of latent syphilis, and useful to show if the virus should be regarded as probably eliminated, or as still remaining in the body. The treatment may thus throw light on the nature of obscure pains, glandular enlargements, loss of hair, etc., which could previously not be traced to syphilis with certainty. This test action is suggested by Güntz to depend on the action of sulphur in aiding the excretion of mercury from the tissues. According to this theory the mercury is stored in the tissues in the form of an albuminate, and prevents the appearance of syphilitic manifestations; the sulphur treatment accelerates the albuminous catabolism, causing therefore the breaking up and excretion of the albuminates of mercury; the removal of the mercury then allows the syphilis to manifest itself again.

Aachen is open all the year round, but there is a summer season, from April 15 to October 15, and a winter season from November to April.

Access: In 11 to 12½ hours from London by Ostend or Calais and Verviers.

Accommodation: Very good. The patients can live in hotels or in the bath-houses themselves, the latter being very convenient during bad weather.

Doctors: Brandis, Mayer, Schumacher, Beissel, Müller, Schuster, etc.

Burtscheid is practically a suburb on the south-east of Aachen, and the springs are similar, but somewhat poorer in sulphur, whilst the temperature of the 'Kochbrunnen' is as high as 162° F., and there is a spring still slightly hotter (the 'Schwertbadquelle.)'

Aix-les-Bains, or Aix in Savoy (France).—This famous spa, the 'Aquæ Gratianæ,' or 'Aquæ Domitianæ,' or 'Aquæ Allobrogum' of the Romans, is situated at an altitude of 870 feet, in the beautiful country of the Savoy Alps, to the east of the picturesque lake of Bourget. The climate is warm and moist and uniform.

The two chief springs, the 'sulphur spring' and the 'Paul's spring' (the latter more usually called the 'alum spring,' though it contains no alum), are nearly devoid of solid constituents, but have a temperature of 109·5° to 112° F. and are very abundant; they are fairly rich in glairine and organic matter, and contain sufficient sulphuretted hydrogen to give them the characteristic smell. The waters of these two springs are chiefly used for baths, but the 'alum spring' is likewise used for drinking.

For internal use, however, the stronger cold sulphur water of Challes (p. 240), containing minute quantities of the iodide and bromide of sodium, is chiefly used at Aix. Challes is near Chambéry, and its water is brought every day to Aix, and can be obtained at all the chemists in the town. At Marlioz, about ten minutes' walk to the south of Aix, is another cold sulphur spring, which is chiefly employed in chronic laryngeal and bronchial catarrhs in adults, and in tendency to bronchitis in delicate children. The inhalation and spray rooms at Marlioz have recently been renewed.

The waters and the various methods of treatment employed at Aix-les-Bains are of service in cases in which indifferent thermal waters are of use, for chronic gouty and rheumatic manifestations, muscular rheumatism, sciatica, neuralgias, neurasthenic conditions in arthritic subjects, chronic cutaneous eruptions, and chronic catarrhal affections of the mucous membranes. For syphilis the principles of treatment are similar to those at Aachen.

Douches and baths are given by preference before

breakfast or in the forenoon, and if mineral water is likewise taken internally, it is usually prescribed before or after the bath, or partly before and partly after, or else before meals. As drinking the waters at Aix plays a part quite secondary to the external treatment by baths and douches, the daily routine is somewhat different to that at the best known German spas, and there is no early morning promenade, with music, as at these spas.

The large bath establishment is the property of the State, and is one of the most efficient of these institutions existing. Poor people are cared for, as well as the rich. There are separate simple thermal baths, and large baths for several people; thermal and cold douches; box vapour baths; 'Berthollet' vapour baths for separate parts of the body; and chambers, known as 'bouillons,' in which the hot mineral water is used to form a vapour bath. There is a special piscina for the treatment of chronic skin affections by prolonged baths after the method of Loèche-les-Bains.

Aix is celebrated for the good results obtained in the stiffness of joints arising from former injuries and from gouty and rheumatic affections. Skilled attendants apply the douches and perform massage according to the doctor's directions. There is one kind of treatment very much employed at Aix, which has received the name of the 'Aix douche' or 'douche-massage'; it consists in the methodical application by two skilled attendants of massage combined with douches, and may be used for the whole body (the head of course excluded), or be specially applied to the diseased part: the sort of massage or rubbing, and the strength and temperature of the douche, must of course be varied according to the individual case, and this treatment may be combined with passive movements of special joints. The 'Aix douche is sometimes followed or preceded by a vapour bath in the adjoining 'bouillon.' At present there is not room at the establishment for patients to rest there after the treatment, but there will soon be additional accommodation for this purpose; in many cases they are carried by porters on chairs from their hotels to the bath and back again to bed after the bath. After a few days' treatment by the 'Aix douche,' an interval of rest, or of days in which a simple thermal bath is taken, is usually prescribed. (For further details of the douche-massage see Forestier, *Le Traitement Thermal d'Aix-les-Bains*, 1895.) The 'douche-massage,' after the method of Aix-les-Bains, has been introduced at several other French spas, and has likewise been adopted with success at Harrogate, Bath, and other places in Great Britain.

The Aix season lasts from April to November, but the bath establishment is open throughout the year. During the main part of the season there is no lack of amusement for patients. An 'after-cure' is often advisable, and for this purpose a stay on the neighbouring Mont-Revard (5,360 feet), up which there is now a rack railway from Aix, or in colder weather, at the less elevated Les Corbières (2,200 feet), may sometimes be recommended. The not very distant health resorts of Les Avants, Caux, and Glion are very convenient, being sunny, possessing a dry soil, and moderately bracing air.

Access: In 19 to 20½ hours from London; *viâ* Paris and Macon.

Accommodation: Very good.

Doctors: Vidal, Brachet, Blanc, Forestier, Monard, Rendall, J. Dardel, Cazalis, Françon, Guilland, Guyenot, etc.

Harrogate (England, Yorkshire). — Harrogate,

situated in a bracing district of Yorkshire, at an elevation from 260 to 600 feet above sea-level, is perhaps the most flourishing English spa, though probably not quite such a centre of fashion as Bath was in the eighteenth century. The lower town is much more sheltered than 'Upper Harrogate,' and not so bracing.

There are about eighty different mineral springs at Harrogate, and the strength and proportion of the constituents vary greatly in the different waters. Most of them are cold muriated springs, containing sulphuretted hydrogen and sulphide of sodium. Of these the 'old sulphur spring' (Royal Pump Room) with about ·07 per mille of the sulphide of sodium and 37 volumes per mille sulphuretted hydrogen, is the one most generally preferred for internal use. It contains likewise 12·7 per mille common salt and ·09 per mille of chloride of barium; the latter substance, of which the amount is about the same as in the Llangammarch waters, is believed to exercise a tonic influence on the heart's contraction, counteracting any depressing effect of the sulphur. The strong sulphur 'Montpellier' spring is said to contain about ·2 per mille sulphide of sodium and no sulphuretted hydrogen. The Starbeck sulphur wells yield mild sulphur waters used for baths.

There are also springs with an appreciable amount of iron; of these the 'Kissingen well' contains ·13 per mille of carbonate of iron, about 1 per cent. of common salt, about 1·2 per mille chloride of calcium, and no chloride of barium, whilst the 'chloride of iron well' contains about ·19 per mille of the chloride of iron, ·16 per mille of the carbonate of iron, about 2·5 per mille of common salt, and about ·07 per mille of barium chloride.

The sulphate of iron water from the 'alum well' in the 'bog field' is interesting on account of its containing, roughly speaking, about one per mille each of ferrous sulphate (protosulphate), ferric sulphate (persulphate), aluminium sulphate, calcium sulphate, and magnesium sulphate. A little marsh gas or carburetted hydrogen occurs in some of the Harrogate springs.

The iron waters are useful in anæmia, but they have not the advantage of containing carbonic acid gas like the foreign spas of Spa, Schwalbach, etc. Though strong sulphur waters should not be given in great anæmia, in some anæmic conditions sulphur treatment is made to precede the iron. It is supposed in these cases that the sulphur stimulates the secretory action of the liver, kidney, and skin, 'clearing out the system,' and preparing it for the beneficial action of the iron. Pharmaceutical preparations of iron may be given in addition to the waters. The common salt of the waters has a favourable influence in anæmic and cachectic conditions and in bronchitic patients. The sulphur and chloride of iron waters are artificially warmed before being taken.

In the New Victoria Baths, Harrogate is provided with recent and elaborate appliances for baths, douches, etc., and there are special attendants for administering the 'Aix douche' after the manner of Aix-les-Bains (*q.v.*). These hydro-therapeutic processes are much employed in the chronic gouty and rheumatic cases treated by the sulphur waters of Harrogate. Cases of commencing osteo-arthritis are more likely to derive benefit from the baths than from the internal treatment.

Sulphur baths are likewise employed in chronic eczema and psoriasis. The claim that a special therapeutic effect is exercised by the sulphur waters in cases of chronic lead and metallic poisoning is not universally admitted.

The Harrogate season is from May to September, but the spa is open also at other times of the year.

Access: From King's Cross Station in about six hours.

Accommodation: Good.

Doctors: G. Oliver, A. S. and J. A. Myrtle, Black, etc.

Askern Spa (England, Yorkshire).—This spa, near Doncaster in Yorkshire, is situated on a large plain, part of which is an imperfectly drained peat bog. It contains a pump room and baths attached to each of its four springs. The waters are alkaline earthy, containing sulphuretted hydrogen gas, and have a yellowish tinge, probably due to their origin in a peat bog.

Their action is diuretic, and they are usually taken cold for internal use to the amount of about half a pint, two or three times a day. For external use, the waters are artificially warmed.

They are employed in some forms of dyspepsia, and in chronic gouty and rheumatic affections.

Access: From King's Cross Station in about four hours.

Accommodation: Good.

Doctors: Hindle, Johnston.

Llandrindod Wells (Wales, Radnorshire).—Llandrindod (altitude 700 feet) possesses muriated waters, sulphur waters, and weak chalybeate waters. The bracing air of the locality contributes much to the good effect derived at the spa, especially to those 'run down' by town life. Llandrindod, although it possessed a local reputation for a long time, has only recently developed into a flourishing health resort. It lies in the centre of an elevated plateau, and though not shut in, is protected towards the east by Radnor Forest.

The muriated waters of Llandrindod (3·4 to 4·8 per mille common salt, 1 to 1·4 chloride of calcium, ·04 to ·7 chloride of magnesium) are slightly laxative, and are useful in atonic cases of dyspepsia, in constipation, in some cases of chronic rheumatism and gout, gouty glycosuria and commencing cirrhosis of the liver; they are not given in intestinal and vesical irritability and in kidney disease. In rheumatic and gouty patients their use is often combined with that of the sulphur waters, and in atonic conditions of the alimentary canal with that of the astringent iron waters.

The sulphur waters are weakly muriated springs, containing apparently 1 to 14 volumes per mille of sulphuretted hydrogen gas. They are useful in some irritable conditions of the alimentary tract, with a tendency to diarrhœa, in some chronic affections of the bladder, and in various cutaneous affections. The sulphur water can be useful, associated with the iron water, in scrofulous affections.

The iron in the chalybeate spring, in spite of its small amount (·018 per mille of the carbonate of iron), is said to be of use in anæmic conditions. The water of this spring contains also about 4 per mille common salt and 1 per mille chloride of calcium. The muriated water is, moreover, sometimes given simultaneously with it.

There are arrangements for baths and douches at Llandrindod.

The season is from May to October. A good locality for an after-cure is Lake Vyrnwy Hotel, about 1,000 feet above sea-level, six hours distant from Llandrindod.

Access: By railway, in about six hours from London.

Accommodation: Good.

Doctors: Bowen Davies, F. H. Roberts, etc.

Builth Wells in Brecknockshire has waters similar to those of Llandrindod. Some of those who take the waters stay at Builth, about 1½ mile from the wells. [Dr. Bennett at Builth.]

Llanwrtyd Wells (Wales, Brecknockshire).—Llanwrtyd Wells lies at an elevation of 800 feet above the sea-level, has a bracing climate, and is considered to rank second to Llandrindod amongst Welsh spas. It possesses sulphur waters said to contain 36 volumes per mille sulphuretted hydrogen, and weak chalybeate springs, containing ·011 carbonate of iron.

The muriated water of Builth may likewise be obtained if required.

The season lasts from May to September.

Access: From Euston Station in about 6½ hours.

Accommodation: Good.

Doctor: W. E. S. Stanley.

Strathpeffer (Scotland, Ross-shire).—Strathpeffer (altitude, 200 feet) lies sheltered in a valley, so that the climate is mild. The strongest of its three cold sulphur wells is said to contain about ·02 per mille sulphide of potassium, ·007 sulphide of sodium, and about 40 volumes per mille of sulphuretted hydrogen. The chalybeate 'Saints' Well' is said to contain about ·035 per mille carbonate of iron.

Patients are chiefly treated here for chronic gouty, rheumatic, and dyspeptic troubles, and for chronic cutaneous affections.

Although the internal use of the waters takes the first place at Strathpeffer, various kinds of baths, including sulphur, brine and peat baths are employed, as well as other hydrotherapeutic treatment. The ordinary time for drinking the water is in the morning at eight o'clock and half-past eleven. The spa is open all the year, but the season is from May to October.

Access: From Euston Station (London) in 17 hours or more.

Accommodation: Good.

Doctors: R. F. and J. T. Fox, W. Bruce (Dingwall), Duncan, Mackay.

Moffatt (Scotland, Dumfriesshire).—Moffatt (altitude 400 feet) possesses weak sulphur and chalybeate springs. The situation is beautiful, and the waters, which are obtained some distance off, are found useful in cases of slight anæmia and debility, chronic cutaneous affections, etc.

Access: By train *viâ* Carlisle in about eight hours.

Accommodation: Satisfactory.

Doctors: Grange, Munro, Rutherford, etc.

Lisdoonvarna (Ireland, County Clare).—Lisdoonvarna, the most popular spa in Ireland, possesses cold sulphur (5·5 volumes per mille of sulphuretted hydrogen gas) and weak chalybeate springs. Chronic gouty, rheumatic, and dyspeptic troubles are treated, and some cutaneous affections. The climate is bracing. The season lasts from June to October.

Dr. E. D. Mapother (*Papers on Dermatology*, 1889, p. 91) points out that the benefit derived from a stay at Lisdoonvarna does not seem to be due solely to moderation in diet, for poor people, who are always forced to live moderately, likewise derive benefit from the place.

Doctors: F. Forster, Westropp.

Swanlinbar (or Swanlibar) a small village in County Cavan, Ireland, possesses cold sulphur springs, which were fashionable in former days.

Lucan (County Kildare), a short distance to the west of Dublin, contains weak, cold sulphur waters, and satisfactory accommodation. It was a popular spa at the beginning of the present century. *Leixlip Spa*, close to Lucan, has weakly mineralised waters, with a faint odour of sulphuretted hydrogen. At one time the Leixlip waters were wrongly thought to be chalybeate, and were popular with the people of Dublin. Their temperature was found by Dr. Mapother in 1875 to be 64° F.

Weilbach (Germany, Prussian Province of Hesse-Nassau).—Weilbach is situated at an altitude of 440 feet, between Frankfurt and Wiesbaden, twenty-five minutes' drive from the railway station of Flörsheim. It possesses two mineral springs, the 'Schwefelquelle' and the 'Natron-lithionquelle.'

The Schwefelquelle is a cold, weakly mineralised spring, containing 5·2 parts sulphuretted hydrogen gas in 1,000 parts by volume of water. The water is used for drinking in the case of stout persons with a tendency to hæmorrhoids and enlargement of the liver. It is also used for bathing and inhalation, the latter in catarrh of the respiratory organs.

The water of the Natron-lithionquelle contains 1·2 per mille chloride of sodium, 1·3 per mille bicarbonate of sodium, ·009 per mille bicarbonate of lithium; it may, therefore, be classed amongst the muriated alkaline waters, and is used in gouty conditions, and in some urinary complaints.

The season lasts from May 1 to the end of September. A good deal of research has been carried on at Weilbach to explain the action of sulphurous mineral waters.

Nenndorf (Prussia, Province of Hesse-Nassau).—Bad-Nenndorf (altitude 230 feet), lies near the village of Gross-Nenndorf in a wooded country not far from Hanover. Of its cold sulphur springs the 'Trinkquelle' (1 per mille sulphate of calcium and about 45 per mille volumes of sulphuretted hydrogen) is richest in sulphur, and is the only one used for drinking purposes.

The Rodenberg brine, containing 6 per cent. of common salt, with a trace of sulphuretted hydrogen, is conducted from Soldorf, to be used for bathing purposes at Nenndorf, and may be strengthened, if required, by the addition of 'Mutterlauge.' Sulphurous mud baths ('Schwefel-Moorschlamm-Bäder') and gas inhalations are likewise made use of in some cases.

Patients come to Nenndorf for chronic rheumatism, gout, cutaneous affections, catarrhal conditions of the respiratory organs, etc. The chief season is from May 15 to September 30.

Meinberg (Germany, in the Principality of Lippe-Detmold) 1½ hour's drive from the railway station of Detmold, lies at an altitude of 660 feet on the northern border of the Teutoburger Wald. It possesses several mineral springs, amongst which the most important are a cold earthy sulphur spring (23 c.c. sulphuretted hydrogen in the litre), used for baths, and a cold muriated spring with about 5·5 per mille common salt, which can be used internally. The carbonic acid gas which exudes from the soil is used for gas-baths. Probably, however, the most important treatment at Meinberg consists in its sulphurous mud baths (see p. 22). There are likewise earthy chalybeate springs rich in carbonic acid gas. The season lasts from May 20 to September 10. Scrofula, chronic rheumatism, gout, neuralgias, and gynæcological affections are treated at Meinberg.

Eilsen (Germany, Principality of Lippe-Schaumburg) lies in a valley (altitude 230 feet) protected from the north and east winds. Of the ten cold sulphur springs, the Julianen-brunnen contains the greatest amount of solid constituents (2 per mille sulphate of calcium) and about 49 volumes per mille sulphuretted hydrogen. Sulphur mud baths ('Schwefel-Moorschlamm-Bäder') and gas inhalations are likewise made use of. The nearest railway station (about 1 hour's drive) is Bückeburg, on the Hanover and Minden line. The season is May 30 to September 5.

Bentheim (altitude 800 feet) in Hanover is situated in a forest of oaks near the Dutch border. It possesses a cold earthy spring (1·3 per mille sulphate of calcium, etc.), containing sulphuretted hydrogen, and used for baths in chronic rheumatic affections, etc., often in connection with hydro-therapeutic treatment or massage. The waters can be used for inhalation in catarrh of the respiratory passages.

Langensalza in Thuringia (Prussian Province of Saxony), at an altitude of 660 feet, possesses several cold sulphur springs, of which the strongest contains as much as 47 per mille volumes of sulphuretted hydrogen. The establishment is twenty minutes' drive from the railway station.

Wipfeld (Bavaria).—Near it is the Ludwigsbad, in a protected position, 715 feet above sea-level. The 'Ludwigsquelle,' a cold, earthy sulphur spring (1 per mille sulphate of calcium, 25 per mille volumes sulphuretted hydrogen), is used for drinking and in the preparation of sulphurous mud baths ('Schwefel-Moorschlamm-Bäder'). There are likewise a weak earthy spring, and two weak chalybeate ones.

Kainzenbad or **Kanïtzerbad**, near Partenkirchen, in Bavaria, close to the Tyrolese border, lies at an altitude of 2,620 feet, about 3¾ hours' drive from the railway station of Murnau. Besides the 'Gutiquelle,' a cold sulphuretted hydrogen spring, there are weak alkaline springs containing iodine, and a chalybeate spring.

Treatment by milk and whey is employed. There is likewise accommodation to be had at the Alm-am-Eck, 3,440 feet above sea-level.

Abbach (altitude 1,140 feet) in Bavaria, about half an hour's journey by railway from Regensburg, possesses a weakly mineralised alkaline earthy spring, containing sulphuretted hydrogen. It was known as far back as the thirteenth century, and possesses local reputation in the treatment of hæmorrhoids, etc.

Langenbrücken (Germany, in the Grand Duchy of Baden), at an elevation of about 440 feet, is a station on the railway between Heidelberg and Karlsruhe. It contains weak cold sulphated sulphur springs used for hæmorrhoidal conditions, chronic catarrhs of the respiratory organs, and in the form of hot baths, douches, etc. for chronic rheumatic affections.

Reutlingen, in Würtemberg (altitude 1,110 feet), on the Echaz, a railway station nine miles east of Tübingen, possesses cold sulphur waters containing minute quantities of the bicarbonates of sodium and magnesium.

Bad Boll (Würtemburg) is prettily situated in the Filsthal, at an altitude of 1,340 feet, about four and a half miles to the south of the railway station of Goeppingen. Its sulphur water was already known in the sixteenth century.

Other German cold sulphur waters are those of SEBASTIANSWEILER (altitude 1,570 feet), in Würtemberg; HECHINGEN (altitude 1,540 feet), and TENNSTEDT (altitude 700 feet), in Prussia; BAD HÖHENSTADT (altitude 1,120 feet), in Lower Bavaria. LANDECK (see p. 63), in Prussian Silesia, has been placed in the indifferent thermal group.

Baden in Austria.—Baden, near Vienna (altitude 700 feet), pleasantly situated at the entrance of the Helenenthal, is very much frequented as a summer resort by the Viennese. The earthy thermal sulphur waters, known already to the Romans, have a temperature of 80·6° to 96° F., and are used more for baths than for drinking.

There are large thermal baths for several persons, separate thermal baths, mud baths (local and general), and arrangements for hydro-therapeutic processes. There are, likewise, swimming baths, supplied with the mineral water.

Chronic gouty and rheumatic articular affections, 'muscular rheumatism,' scrofula, and chronic skin eruptions are amongst the conditions treated here. When used internally (as in some cases of chronic bronchial and gastric catarrh), the water is mixed with milk or whey, or with the mineral water of another place. The chief season lasts from May 15 to October 15, but the baths are open during the whole year.

Access: From London with about 33 hours' travelling; from Vienna by train in an hour.

Accommodation: Good.

Doctors: Barth, J. Schwarz, Carl Schwarz, and many others.

Altenburg (or Deutsch-Altenburg), in Lower Austria, near Pressburg, contains a weak thermal sulphur spring, having a local reputation in cases of chronic cutaneous eruptions, etc. This bath (altitude 490 feet) was formerly called Hofbad, and was very famous. In Roman times it was known as 'Thermæ Pannoniæ.'

Innichen, in the Austrian Tyrol, at an altitude of 4,370 feet, is beautifully situated amidst forests in a branch of the Puster Thal, half an hour from the railway station of Innichen. There are two cold sulphur springs and a chalybeate one.

Alt-Prags, beautifully situated at an altitude of

4,500 feet in the Pragser Thal (Austrian Tyrol), is one and a half hour distant from the railway station of Niederdorf. It possesses a weak sulphur spring used for bathing.

In the Austrian Tyrol there are also cold sulphur waters at LÄNGENFELD (altitude about 3,810 feet), in the Oetz Valley, and at LADIS (altitude 3,930 feet), near Landeck.

Hercules Bad (Hercules-fürdo), near MEHADIA, in Hungary (altitude 570 feet), lies in the romantic Czörna Valley, three miles from the Danube, and on the railway between Orsova and Temesvar. Its thermal waters (known from Roman times), with temperatures from 70° to 133° F., are mostly muriated sulphurous ones; they have been compared to those of Aix-la-Chapelle, and like them are employed internally and externally, but chiefly externally. The sulphur is contained in the form of sulphuretted hydrogen, one spring having about 42 per mille volumes, it is said; but the 'Hercules spring' is quite free from it, and has therefore been already mentioned in the chapter on muriated waters. The same affections are treated as at Aix-la-Chapelle.

The situation at the foot of the Karpathian Mountains is very beautiful and much appreciated by the people of south-eastern Europe. The accommodation is good. Season: May to the end of September.

Doctors: Popovich, etc.

Other Hungarian thermal sulphur waters are those of PYSTJAN (135°–146° F.), TRENCZIN-TÖPLITZ, or TEPLITZ-TRENTSCHIN (99°–104° F.), HAJO, near GROSSWARDEIN (99°–111° F.), and HARKANY (143–5° F.). In the waters of Harkany, Karl von Than discovered the inflammable gas, sulphide of carbonyl (COS), which is said to be present over the spring in a quantity sufficient to be ignited.

Buda-Pest (see p. 67) possesses thermal sulphur waters.

Balf, in Hungary, is a village with mild climate, 1 mile from Oedenburg. It possesses two cold muriated alkaline sulphur springs, which are used by patients from the neighbourhood only.

Parad, in Hungary, contains a strong sulphuretted hydrogen spring. The spa has already been noticed amongst sulphate of iron waters. (See p. 196.)

Warasdin-Teplitz, or **Warasdin Töplitz**, in Croatia, three hours from the railway station of Csakathurn, lies at an altitude of 920 feet, in a pleasant position sheltered from the north. Its thermal sulphur waters (temperature 136·4° F.) are said to have been known to the Romans as the 'Aquæ Jasæ.' Their total of solid constituents is ·77 per mille.

Baden in Switzerland (Canton Aargau).—The spa (altitude about 1,230 feet), and the somewhat higher situated old-fashioned town of Baden, lie in a beautiful valley on the banks of the River Limmat. The position of the place is a fairly sheltered one, the climate is mild, and is influenced by the extensive forests surrounding the place. Its thermal weak sulphurous waters were known already to the Romans, and were famous in the Middle Ages, when the Papal Secretary, Poggio Bracciolini, described the gay nature of the spa-life here (1416).

The average temperature of the waters is 118·4° F.; they smell of sulphuretted hydrogen, and contain a certain amount of the sulphates and chlorides of calcium and sodium, and it is said that a mentionable quantity of arsenic can be detected in them. Owing to the earthy constituents the water of Baden is not much taken internally, but when this is advisable it may be mixed, as in some hæmorrhoidal cases, when a laxative effect is required, with the neighbouring Birmensdorfer bitter water, or in other cases, a little bicarbonate of sodium may be added.

The different hotels have their own baths, but there is likewise a separate bath establishment, which is used by certified poor patients of different countries. The patients treated at Baden include many with stiff joints of a chronic rheumatic or gouty nature, or resulting from injury, or previous peripheral neuritis. Others come there for sciatica, lumbago, 'muscular rheumatism,' and various affections on a gouty basis. The baths are usually given at a temperature of about 93° F., and the time preferred is before breakfast. When a more stimulating effect is desired, salt from the neighbouring Rheinfelden brine can be added. Massage is much employed for the joint affections, and for cases of sciatica and muscular rheumatism. Inhalation rooms are provided for use in chronic affections of the air passages.

The neighbourhood affords excellent ground for a 'terrain-cur' after Oertel's views. The season at Baden is from the middle of May to the end of September, but the spa remains open all the year. Owing to the waters being little drunk, there is no 'Kur-Musik' in the early morning before breakfast, and in this respect the ordinary daily routine differs somewhat from that at most of the well-known German spas.

Access: In about 21 hours, *viâ* Bâle.

Accommodation: Very good.

Doctors: Minnich, Röthlisberger, Borsinger, Keller, etc.

Schinznach (Switzerland, Canton Aargau).—Schinznach, a station on the railway from Zurich to Aargau, lies at an altitude of about 1,140 feet, in the pleasant valley of the Aar. The establishment is situated in grounds of its own, distinct from the village, and is likewise known as the 'Habsburger Bad,' from the ruins of Habsburg, which crown the neighbouring Wülpelsberg (1,680 feet.)

The Schinznach spring yields thermal strong sulphurous waters (37 per mille volumes of sulphuretted hydrogen), with a temperature varying from 82·4° to 95° F., and containing about 1 per mille sulphate of calcium. The bath establishment affords good lodging, as well as baths for the patients; it is fitted up for the mineral water baths, vapour baths, and ordinary baths; also with apparatus for nasal and local douches, and for inhalation of the pulverised mineral water, and the gases given off.

The mineral waters of Schinznach are used both internally and externally. Baths of long duration (1½ to 2 hours) are often prescribed, and the water has sometimes to be heated one or two degrees for bathing. If taken internally it is recommended that as a rule the dose be taken before the bath.

The affections treated here are chronic eczema and other chronic skin eruptions (for which the spa has a special reputation), chronic gouty and rheumatic complaints, leucorrhœa, chronic catarrhal conditions of the respiratory organs, caries of bone, scrofula, and rickets. Nasal douches, sprays, and inhalations are employed in naso-pharyngeal catarrh, bronchitis, asthma, and emphysema. In some scrofulous and cutaneous affections the muriated water of the neighbouring Wildegg (see p. 122), containing small quantities of iodides and bromides, is recommended for internal use. The season lasts from May 15 to the end of September.

Doctors: Amsler (of Wildegg), von Tymowsky, Hemmann.

Lavey (Switzerland, Canton of Vaud), 1¼ mile from the railway station of Saint Maurice, possesses weak thermal sulphur waters (temperature 92° to 118°), containing about 3 per mille volumes sulphuretted hydrogen gas, and about 1 per mille solids (sulphate and chloride of sodium). The waters are used for drinking, bathing, and inhalation. For baths the 'Eau mère' of Bex is added to the water, and sometimes also for drinking. Baths of sand from the Rhone are employed at a temperature of 149° F. Lavey lies in the Rhone Valley at an elevation of 1,350 feet.

Yverdon (Switzerland, Canton Vaud) is situated at the southern extremity of the Lake of Neuchâtel, on the railway between Lausanne and Neuchâtel (altitude 1,420 feet). Its sulphur water (temperature 75° F.) is feebly mineralised, and contains 3·4 per mille volumes of sulphuretted hydrogen.

Lenk (Switzerland) lies in a sheltered position at an altitude of 3,680 feet, in the beautiful Obersimmenthal of Canton Bern. By misprints in the word it has sometimes been confused with Leuk (see Loèche-les-Bains) in Canton Valais. Lenk possesses two cold sulphur springs, of which the strongest, the 'Balmquelle,' contains 44·5 c.c. of sulphuretted hydrogen gas in the litre. There is likewise an earthy weak chalybeate spring (·01 per mille bicarbonate of iron). The railway station of Thun is 8½ miles distant by diligence. The season is from June 15 to September 30.

Gurnigel (Switzerland, Canton Bern) lies near the Stockhorn at an elevation of about 3,550 feet above sea-level. Its cold sulphur spring, 'Schwarzbruennli,' contains 1·3 per mille sulphate of calcium, ·04 per mille sulphide of calcium, and ·01 per mille sulphide of magnesium, with 24 per mille volumes of sulphuretted hydrogen gas. The 'Stockquelle' water contains less sulphuretted hydrogen. The climate is sunny and bracing; there are beautiful walks in the adjacent pine forests, and the accommodation is excellent. The railway station of Bern is 4½ hours distant.

Heustrich (Switzerland, Canton of Bern) lies on the eastern slope of the Niesen at an elevation of 2,000 feet, 2 hours by carriage from the railway station of Thun. Its cold sulphur waters, having a total mineralisation of about 1 per mille, contain sulphide of sodium (·03 per mille) and sulphuretted hydrogen (11 per mille volumes) in association with small quantities of the bicarbonate (·6 per mille) and sulphate of sodium. They are used in chronic catarrhal conditions of the respiratory passages, etc. Accommodation at Heustrich is good.

Schimberg (altitude 4,670 feet), in the Canton of Lucern, possesses similar waters to those of Heustrich, but containing rather less sulphur. The establishment is situated on the western slope of the Schimberg mountain, by which it is protected from north-east winds, though the south-west and south winds are sometimes violent. For baths the water of another spring (which is termed chalybeate) is employed.

Lostorf (Switzerland, Canton Solothurn) lies at an elevation of 1,640 feet, on the southern declivity of the Jura, and possesses a cold muriated sulphur spring (3 per mille common salt), and two other springs, of which one is similar to the first but weaker, and the other has a weak earthy mineralisation.

Alveneu, or **Alvaneu** (Switzerland, Grisons), eleven miles from the station of Chur, lies at an elevation of

3,150 feet above the sea level. It possesses a cold earthy sulphur spring (containing about 1 per mille sulphate of calcium and very little sulphuretted hydrogen) used in chronic rheumatic and gouty joint affections, catarrh of the respiratory organs, etc. In the neighbourhood are the sulphated alkaline chalybeate springs of St. Peter at TIEFENKASTEN (2·2 per mille sulphate of sodium, 1·7 per mille bicarbonate of calcium, and ·029 per mille bicarbonate of iron), and of St. Donatus at SOLIS; the latter contains a small amount of iodide (·001) and bromide (·002) of sodium. The season is from June 15 to September 25.

Le Prese in Switzerland (Canton Grisons) is a summer resort on the Lago di Poschiavo, 3,160 feet above the sea level, half an hour's drive from Poschiavo, and about six hours' drive from Samaden. The cold sulphur waters are feebly mineralised, and contain ·6 c.c. sulphuretted hydrogen in the litre; they are used for baths and for drinking. The bath arrangements and accommodation are satisfactory. The season is from the commencement of June to the end of September.

Serneus (Switzerland) in the Grisons has a cold sulphur spring with about 9 per mille volumes of sulphuretted hydrogen gas. The establishment lies at an altitude of about 3,240 feet in the Landquart valley on the branch railway from Landquart to Davos.

Stachelberg (Switzerland, Canton Glarus), near the railway station of Linththal, possesses a cold sulphur spring, containing very little sulphuretted hydrogen, but ·04 per mille sulphide of sodium. By its climate and beautiful position in the Toedi district, 2,050 feet above sea level, it offers considerable advantages.

Other Swiss cold sulphur waters are those of MONTBARRY (altitude 2,460 feet) in Canton Freiburg, FLUEHLI

IM ENTLEBUCH (altitude 2,930 feet) in Canton Lucern, and RIETBAD (altitude 2,790 feet) in the Toggenburg district of Canton St. Gall.

Eaux Chaudes (France, Basses-Pyrénées).—The village is situated at the southern end of the narrow and romantic Ossau valley (altitude 2,050 feet), four miles from the railway station of Laruns and five miles from the spa of Eaux Bonnes. Its mountainous position causes considerable daily fluctuations in the temperature. The thermal springs have a temperature of 92°-97° F., and are similar in their mineral constituents to those of Eaux Bonnes, but contain less sulphur (sulphide of sodium ·0088 ; total solids ·33 per mille).

The waters (contrary to those of the neighbouring Eaux Bonnes) are employed chiefly for baths and douches. The different springs from old tradition have reputations for different affections ; the 'Source Clot' for arthritic affections, the 'Source Esquirette' for uterine troubles, etc., the 'Source Rey' for nervous disorders in rheumatic subjects, and the 'Source Baudot' for catarrhs of the respiratory organs. The waters of Eaux Chaudes have a less excitant action than the hotter Pyrenean sulphur waters, but have, it is said, a tendency to produce hyperæmia of the pelvic organs, and thereby aid in the re-establishment of the menses in chlorotic girls. The season lasts from June 1 to October 1.

Barèges, France (Hautes-Pyrénées).—This celebrated spa lies in the narrow valley of the Bastan, at an elevation of 4,200 feet above the sea level ; in summer warm clothing should be brought, and the place is hardly habitable during winter. The waters are thermal sulphurous (temperature 81°-111° F.), and do not whiten on exposure to the air as do those of Luchon ;

they contain an organic substance which forms a scum on the surface, and was named by Longchamp 'Barégine' after this spa. In courses of baths the tepid ones are used to begin with, and then gradually the hotter ones. The hot waters have a powerful nervous excitant action.

Barèges has a very great reputation in the treatment of old gun-shot and other wounds, painful cicatrices, and chronic joint affections; there is a large hospital for officers and soldiers. Chronic eczema and psoriasis are said to be at least temporarily benefited. Barèges is also much resorted to by sufferers from syphilis. The spa became famous in 1675 when the Duc du Maine, natural son of Louis XIV., was treated with good result for a tuberculous affection. Internally used the waters sometimes give rise to nausea and diarrhœa; they are less used for drinking than for bathing. The 'Tambour' spring (temperature 115° F.), which contains ·04 per mille sulphide of sodium, is the only one used internally, and is taken in small doses, often mixed with milk or whey.

The season lasts from June 15 to September 15.

Access: 2½ hours by carriage from the railway station of Pierrefitte (see under Cauterets).

Accommodation: Moderate.

Doctors: Armieux, Bétous, Grimaud, etc.

Barzum, close to Barèges, is a spring with water similar to that of Barèges. In 1881 the water of Barzum was conveyed by a conduit 4½ miles to Luz, a village situated at a lower altitude than Barèges, and only 1½ hour's drive from the railway station of Pierrefitte.

Saint Sauveur (France, Department of Hautes-Pyrénées).—The village is situated in the valley of the

Gavarnie part of the Gave, which joins the Cauterets part of the Gave at Pierrefitte. It is about 2 hours' drive from the railway station of Pierrefitte, and lies at an elevation of 2,500 feet above the sea.

The 'Source des Bains' or 'des Dames' (temperature 94°) supplies the bath establishment and contains about ·02 per mille sulphide of sodium. The other spring, the 'Source de la Hontalade' (temperature 86°), a few minutes' walk from the village, contains slightly less sulphide, and has a special reputation in cases of gastralgia, like the Mahourat spring has at Cauterets.

St. Sauveur may be called a 'ladies' spa,' and is mostly used for gynæcological affections and functional nervous disorders.

The season is from June 1 to October 1.

Bagnères-de-Bigorre (France, Hautes-Pyrénées) possesses sulphur waters in addition to its other waters. See the chapter on Earthy Waters, p. 256.

Ax-les-Thermes (France, Department Ariège), near Tarascon, possesses about sixty thermal sulphur springs, varying in temperature from 66° to 163·5° F. The town is picturesquely situated at an altitude of about 2,340 feet at the southern extremity of the Ariège valley. The climate is somewhat changeable, and it becomes very cold in the evenings. The waters contain about ·02 per mille of sulphide of sodium, and are employed in chronic rheumatism, 'torpid' scrofulous affections, chronic skin eruptions, and chronic bronchitis.

The Ax waters, termed 'degenerated sulphur' waters, in which the sulphide of sodium has been converted into the hyposulphite and sulphate, are said to exercise a sedative action, whereas, a stimulating action is exerted by the group of waters in which the sulphur exists as the sulphide of sodium.

The season is from May 15 to October 30.

Access: Ax is the terminus of a railway from Toulouse.

Accommodation: Good.

Doctors: Auphan, Bonnans, Dresch, Fugairon, Palenc, Pujol.

Amélie-les-Bains (called Arles-les-Bains until renamed by Louis Philippe in honour of his wife), France (Pyrénées-Orientales).—This well-known spa is situated at an altitude of 920 feet in a valley shut in by high mountains, and although, owing to this, the sun only shines for a very short time in the day, the climate in winter is mild and dry. The spa is open all the year round, but the chief season is from May to the end of October. The east wind is sometimes disagreeable in spring.

The various springs yield alkaline sulphur waters, which have a temperature of 92°–145° F., and are rich in glairine and organic matter. The action of these waters is somewhat excitant, and they are contra-indicated in patients of erethic constitution. The Romans made use of the waters, and one of the two bath establishments is erected on the foundations of ancient Roman thermæ.

The baths are used for skin affections, chronic rheumatism, pain in old wounds, etc., and treatment by various hydro-therapeutic appliances may be added. The waters are inhaled in affections of the respiratory system, for which the mild climate of the spa is suitable—notably during the winter months. Drinking the waters is recommended in some disorders of the liver and digestive system. There is a military hospital with baths of its own.

Access: Railway to Perpignan and on to Céret; thence a six miles' drive to Amélie-les-Bains.

Accommodation: Good.

Doctors: Arnal, Genieys, Lamarchand, Picard, Pujade.

Le Vernet (France, Pyrénées-Orientales), about five miles from the railway station of Prades, lies at an altitude of about 2,060 feet at the foot of the Canigou peak. It possesses thermal sulphur springs (90° to 136·5° F.), containing about ·04 per mille sulphide of sodium ; there is also one cold sulphur spring. The waters are employed for drinking, bathing, and inhalation. Patients resort to Le Vernet for chronic affections of the respiratory organs, chronic rheumatism, cutaneous eruptions, etc.

La Preste (France, Pyrénées-Orientales) is a village situated at an altitude of 3,660 feet, about 20 miles from Amélie-les-Bains. Its thermal sulphur waters (temperature 88° to 111° F.) decompose rapidly on exposure to the air.

Olette (France, Pyrénées-Orientales) lies in the narrow valley of the Tet. It possesses thermal sulphur springs (temperature 90° to 172° F.) containing about ·03 per mille sulphide of sodium. The bath establishment is situated about two miles from the village at an elevation of 2,300 feet above the sea.

Molitg (France, Pyrénées-Orientales) lies at an altitude of about 2,000 feet, in the valley of the Tet, about six miles from Le Vernet. It possesses 12 thermal springs having temperatures varying from 89° to 100·5° F., and containing from ·003 to ·018 per mille sulphide of sodium.

Les Escaldas (France, Department of Pyrenées-Orientales) is situated close to the Spanish frontier on a plateau, about 4,430 feet above the sea-level. The waters are thermal, and contain sulphide of sodium. The temperature of the 'Grande Source' is 109·5° F.

Uriage (France, Department of the Isère).—Uriage, about 8 miles from Grenoble, lies at an altitude of 1,350 feet in a beautiful valley of the Dauphiné Alps. The thermal muriated sulphur spring (temp. 81° F.) contains 6 per mille common salt, about 1 per mille each of sulphate of sodium and sulphate of calcium, ·6 per mille sulphate of magnesium, ·0021 per mille arseniate of sodium, and 7 volumes per mille of sulphuretted hydrogen gas. In doses of from four to six glasses it has a laxative effect.

The water is used in scrofula, chronic cutaneous affections, rheumatism and uterine affections. Both baths and douches of the mineral water are employed, and massage is combined with the douches as in the 'douche-massage' of Aix-les-Bains. The water of a chalybeate spring is likewise made use of in some cases. The season lasts from May 15 to October 15. The nearest railway station, Gières, is about an hour's drive.

Accommodation: Good.

Doctors: Doyon, etc.

Allevard (France, Department Isère) possesses cold muriated sulphur waters. The waters are used for drinking and inhalation in cases of chronic catarrh of the respiratory organs, and for prolonged baths in cases of chronic skin eruptions. A 'whey-cure' is employed, and whey baths may also be taken, which are said to possess a sedative action. The spa is prettily situated at an elevation of about 1,400 feet above the sea-level, and the thermal establishment was rebuilt in 1893, and provided with new appliances. The season lasts from May 15 to October 1. Allevard is about six miles from the railway station of Goncelin (omnibus in 1½ hour).

Challes (France, Department Savoie), about three miles by tram from the railway station of Chambéry, lies at an altitude of 880 feet, in a position sheltered from north winds. It possesses strong cold sulphur waters, which are supplied for internal use at Aix-les-Bains (*q.v.*), and are now also employed at the place itself. The total mineralisation of the water is 1·3 per mille, and the quantity of sulphur, reckoned as sulphide of sodium, is ·5 per mille; small quantities of iodide of sodium (·01 per mille) and bromide of sodium (·003 per mille) are likewise contained.

The water is used for drinking and for inhalation of the spray in cases of chronic catarrhal conditions of the throat and pharynx, in ozœna, adenoid vegetations, and chronic bronchitis; also in cases of scrofulous children, syphilitic cachexia, etc. For baths the water is diluted.

Doctor: Raugé.

Gréoulx (France, Department of Basses-Alpes) lies at an altitude of 1,140 feet, about 2 hours distant from the railway station of Manosque. Its thermal sulphur waters are used for baths at their natural temperature of 97° F.

Bagnols (France, Department of Lozère).—Lies in the narrow valley of the river Lot, at an altitude of about 2,600 feet, and is 23 miles distant from the railway station of Villefort. It possesses thermal, weakly-mineralised springs (temperature 95° to 106° F.), containg about 1·7 vols. per mille of sulphuretted hydrogen. The waters are employed for drinking, for hot baths, and for inhalation.

Good results are claimed at Bagnols in chronic rheumatism and skin diseases, and also in some chronic cardiac affections. At all events, it was long ago shown at this spa that thermal baths might be taken without harm, and in some cases with advantage by patients suffering from chronic [rheumatic] affections of the cardiac valves, without loss of compensation. [See J. E. Dufresse de Chassaigne, 'Mémoire sur le traitement et la guérison de l'anévrysme rhumatismal du cœur (endocardite rhumatismale chronique) sous l'influence de l'usage des eaux thermales de Bagnols.' Angoulême, 1859.]

Saint-Honoré (France, Department of Nièvre).—Saint-Honoré lies in a pleasant country, at an altitude of 990 feet, about six miles from the railway station of Vaudenesse and 32 miles from Nevers. Its tepid waters (temperature 72° to 88° F.) were known to the Romans, and contain a little sulphuretted hydrogen gas and traces of arsenic (said to be ·0012 per mille in the 'Source Crevasse'). They are used (drinking, bathing, and inhalation) in chronic affections of the respiratory organs, scrofula, and chronic cutaneous eruptions.

Season: May 15 to October 1.

Enghien (France, Department Seine-et-Oise) is a small town (altitude 160 feet) close to Paris. It possesses cold sulphur springs and well-arranged establishments. The waters are used for drinking, bathing, douches, and inhalation.

Pierrefonds (France, Department of Oise).—This little town, celebrated for its feudal castle, rebuilt by Viollet-le-Duc for Napoleon III., stands on the edge of a small lake below the hill on which the castle rises, at the southern border of the forest of Compiègne. Its cold sulphur spring contains ·015 per mille sulphide of calcium, small quantities of earthy salts, and 1·4 volumes per mille sulphuretted hydrogen. The waters are especially used in chronic affections of the respiratory organs. There is also a cold chalybeate spring, said to contain ·139 per mille bicarbonate and crenate of iron, small quantities of earthy salts, and traces of manganesium and arsenic.

The season is from June 1 to September 30.

Pietrapola, in the island of Corsica, is picturesquely situated in a mountainous region, and contains thermal sulphur springs (temperature 90° to 137° F.)

Puzzichello, in the island of Corsica, at a low elevation, possesses cold sulphur springs, which have a reputation in the treatment of cutaneous affections.

Guagno, in the western part of Corsica, about 40 miles north of Ajaccio, possesses thermal sulphur springs (temperature 126° F.) and a military hospital. The waters are employed for skin diseases, old gun-shot wounds, etc., as at Barèges.

Acqui, in North Italy (Province of Alessandria).—Acqui (altitude 450 feet), in North Italy, 21 miles south-east of Alessandria, on the railway to Savona, possesses hot muriated sulphur springs, known already in Pliny's time as 'Aquæ Statiellæ.' The climate is moist and changeable, and hence patients should be provided with warm clothing Of the eight springs the most important is La Bollente (1·5 per mille common salt), which emerges at 158° F., and in the different chambers has a temperature of 118° to 124° F. The temperature of the other springs is 102° to 142° F.

Acqui has a reputation for gouty and, especially, rheumatic joint complaints, and those resulting from injuries, also for neuroses and some skin affections. The local application of a hot mud-like substance, which is brought up from the bottom of the well, impregnated with organic matter and with the salts of the mineral waters, plays a chief part in the treatment, and may be compared to the similar treatment at Abano, Battaglia, and Valdieri, and to the mud and peat baths of Dax, Franzensbad, etc. The season is from May 15 to September. 30

Doctors: Schivardi, Alessandri, etc.

Abano (North Italy), one of the Euganean spas. The thermal waters have been described in the Muriated group. (See p. 125.)

Battaglia, in Italy, has often been included amongst the thermal sulphur springs, but its waters contain no sulphur, and are best classed with the indifferent thermal group. (See p. 75.)

Porretta, in Italy (Province of Bologna), lies in the valley of the Reno, amongst the Apennines, at an altitude of 1,100 feet. It is a railway station on the line from Bologna to Pistoja, 37 miles distant from the former. The thermal waters (temperature 91° to 95° F.), known from ancient times, contain 8 per mille common salt (Sorgente Leone), traces of iodides, bromides, and arsenic, a little sulphuretted hydrogen gas, and some of the inflammable carburetted hydrogen or marsh gas. The action of the waters is laxative and diuretic; they are used in cases of hæmorrhoids, 'abdominal plethora,' etc., and in the form of baths for cutaneous affections and chronic rheumatism.

The marsh gas likewise exudes from fissures in the Sasso-Cardo Mountain above the town, and, according to Dr. Macpherson, can be collected in such considerable amounts that it has been at times utilised for lighting the town. The season is from June 30 to September 30.

Valdieri, in North Italy (Piedmont) lies at an altitude of 2,700 feet, in the valley of the Gesso. Its waters have been classed, like those of Battaglia, in the indifferent thermal group (see p. 76).

Acireale, a flourishing town of Sicily, lies at an altitude of 530 feet, near the coast, at the foot to the south-east of Mount Etna. It is used as a winter climatic health resort, and possesses the 'Santa Venera' sulphur wells (70° F.), which contain according to Silvestri (1872) 2·6 per mille common salt, ·01 per mille iodide of sodium, and, in 1,000 volumes, 10 volumes of sulphuretted hydrogen, 95 of carbonic acid, 21 of nitrogen, and 10 of carburetted hydrogen or marsh gas.

Panticosa (PENTICOUSE) in Spain, is situated at an altitude of about 5,400 feet in the Pyrenees, near the French frontier, 12 hours' ride from Eaux Chaudes. The spa lies 5 miles from the village of Panticosa, and has almost the highest situation of any European baths. St. Moritz, in Switzerland, is, however, higher.

The principal spring, chiefly used for drinking, is called 'del Hidalgo,' or the 'liver spring,' and this, as well as the 'Fuente de los Herpes' (*i.e.* 'eruption spring') and the 'Fuente de la Laguna,' may be classed amongst the indifferent thermal waters (temperature 77° to 84·5° F.). The 'Fuente del Estómago,' or 'Stomach spring' (temperature 84·5° F.), is a sulphurous spring containing ·002 per mille sulphide of sodium and some sulphuretted hydrogen gas.

Climate must take a great share in the results obtained at this spa. The waters are especially employed in affections of the respiratory organs, in dyspeptic conditions, and in chronic skin eruptions. The season lasts from June 15 to September 15.

Trillo (Spain, Province of Guadalaxara) lies on the Tagus, 50 miles from Madrid. It possesses thermal muriated chalybeate waters, which smell of sulphuretted hydrogen, and have a temperature of 77° to 86° F. The water is used externally in rheumatism, skin eruptions, etc., and also internally in some cases.

Caratraca, in Spain, is situated in a beautiful country, not far from Malaga, and possesses thermal weakly mineralised sulphur waters (temperature 66° F.), which have a reputation in Spain for skin affections and syphilis.

The thermal sulphur waters of LEDESMA, in Spain (Province of Salamanca), in a fine situation at a considerable elevation above the sea, are much frequented by Spaniards. Other thermal sulphur waters in Spain are those of CORTEGADA, in the Province of Orense (this place possesses also sub-thermal chalybeate springs); CARBALLINO, in the same province; CARBALLO, in the Province of Corunna; ONTANEDA, in the Province of Santander; and ARCHENA, in the Province of Murcia. SANTA AGUEDA, in the Province of Guipuzcoa, in the North of Spain, possesses cold earthy sulphuretted hydrogen waters, and a chalybeate spring.

Caldas-de-Rainha (Portugal, Province of Estremadura), possesses weak thermal muriated springs, containing sulphuretted hydrogen gas (temperature 96° F.). The water is used internally and externally in cases of debility, chronic rheumatism, etc. This spa is the most frequented one of Portugal, and is beautifully situated. There are two hospitals.

Other Portuguese thermal sulphur waters are those of CALDAS-DE-VIZELLA (said, according to Macpherson, to taste like Harrogate waters), and the very hot waters of SAN PEDRO DO SUL (about 152° F.).

Pjatigorsk (Russia) lies at an altitude of 1,640 feet in the Caucasus Mountains, and possesses thermal sulphur springs, having temperatures from 83·5° to 117° F.

Sandefjord, in Norway, is a small town prettily situated on a little 'Fjord' on the North Sea, about four or five hours by railway from Christiania. It is the oldest mineral water station in Norway (the bath-house was built in 1837), and contains cold gaseous muriated sulphur springs, used for drinking and bathing. There are also a chalybeate spring (containing, it is said, 1·29 per mille sulphate of iron and some alum) and a cold weak muriated spring with 4·4 per mille common salt.

Cold and hot sea-water baths are made use of, and a sulphurous slimy material found in the Fjord[1] is employed for rubbing the body, and in the form of hot applications for chronic articular rheumatism, etc. This practice was adopted from that existing at the marine spa of Strömstad in Sweden. Another curious practice is the application of living Jelly Fishes (*Medusa aurita*, *Cyanea capillata*) to produce a sort of counter-irritation of the skin in chronic rheumatic affections, neuralgias, etc. The season is from the beginning of June to September 1.

[1] See Ebbesen and Hörbye. 'The Sulphurous Bath at Sandefjord in Norway.' English translation. Christiania, 1862.

CHAPTER XV

EARTHY OR CALCAREOUS WATERS

These waters (see p. 16) differ much in the proportion of their constituents. Some of them, such as Bath and Loèche-les-Bains, have, for greater convenience, been classed under the simple thermal waters; others, such as Baden in Austria, Baden in Switzerland, and Schinznach, under the sulphur waters.

Some of these waters (see p. 26), owing to their alkaline and astringent nature, act beneficially in digestive troubles, with tendency to attacks of diarrhœa and undue irritability of the mucous membrane. In skin diseases, such as chronic eczema and psoriasis, their action when used in the form of prolonged thermal baths, by soaking and cleansing the skin, is doubtless more important than any special action derived from their solid mineral constituents (see p. 49). Whether in cases of osteomalacia, rickets, and tuberculosis, these waters have any special therapeutic value beyond that of aiding digestion, seems doubtful.

To their diuretic action (see p. 26) waters, such as those of Contrexéville, probably owe part of their repute in cases of urinary gravel, chronic vesical catarrh, etc. The chalk in the waters of Wildungen, etc., appears not in any way to increase the size of urinary concretions, unless in the case of phosphatic calculi they indirectly favour fresh deposits by increasing the alkalinity of the urine. Whether any of these waters have the power which has been claimed for them of inducing the breaking up and expulsion of urinary calculi appears doubtful; such calculi have been known occasionally to undergo spontaneous fracture and expulsion.

Amongst the spas of this group, Wildungen and Contrexéville will be described first, and the rest arranged in geographical order.

Wildungen (Germany, Principality of Waldeck).—Wildungen is picturesquely situated in an open valley at an elevation of about 980 feet above the sea-level, and is fairly sheltered from cold winds. Bad-Wildungen proper is the western portion of the town, and consists nearly entirely of one long street, the 'Brunnen-Allée,' in the villas and hotels of which most of the patients lodge. The neighbouring woods afford delightful walks to those patients for whom open-air exercise is recommended.

At the western end of the Allée is the Georg-Victorquelle, medicinally made use of at least since the sixteenth century; here, during the season, the band plays in the morning, whilst the patients drink their water. The Helenenquelle is situated in the beautiful Helenenthal, about half an hour's walk in a south-westerly direction from Wildungen. The Königsquelle, near the railway station, is the private property of one of the Wildungen doctors. These three springs supply cold gaseous water, containing ·5 to 1·3 per mille of the bicarbonates of calcium and magnesium, and ·018 to ·036 of bicarbonate of iron. The Georg-Victorquelle is the least strongly mineralised (total mineralisation is only about 1·4 per mille), and but for its containing about ·029 per mille bicarbonate of iron, might be classed as a 'table-water.'

The Königsquelle contains the most iron (·036 of the bicarbonate), whilst the Helenenquelle contains ·84 per mille bicarbonate of sodium, and both these springs contain a little over 1 per mille of each of the three salts—bicarbonate of calcium, bicarbonate of magnesium, and common salt.

Besides the above-mentioned three springs there is the (but little used) earthy chalybeate 'Thalquelle,' about two miles distant from the town, and near it is the 'Stahlquelle,' a strong fairly pure chalybeate spring (·07 per mille bicarbonate of iron), rich in carbonic acid gas. For the convenience of patients the waters of the Stahlquelle and the Helenenquelle, as well as milk and whey, are supplied at the Georg-Victorquelle.

The bath-house, which is also a dwelling-house ('Badelogirhaus') for patients who care to live there, is situated close to the Georg-Victorquelle, but is supplied by a separate spring. There is likewise a small bath-house at the other end of the town, attached to the Königsquelle.

The patients who resort to Wildungen nearly all suffer from affections of the urinary system, or at least have symptoms resembling those due to one of these affections. There are patients with vesical calculus, chronic cystitis, pyelitis, enlarged prostate and its results, gonorrhœa, and urethral stricture. Some suffer from uric acid gravel, and some have slight albuminuria, with or without organic change in the kidneys.

The diet in the hotels is regulated to suit the class of cases chiefly met with at Wildungen; beer, mustard, highly-seasoned and rich dishes are hardly to be seen on the tables; patients are especially recommended to observe great moderation in alcoholic drinks and sweet dishes. It must not, however, be supposed that Wildungen is merely resorted to by patients, for its agreeable situation attracts ordinary visitors and tourists, and will probably cause it to be still more esteemed as a summer resort than it is at present.

The waters are taken in the morning before breakfast, often again at noon, before the midday meal, and sometimes once more in the afternoon. The water, in the case of many patients, is best warmed before drinking, though most of the carbonic acid gas must thereby escape. For this purpose troughs of hot water are supplied, and the glasses containing the mineral water are allowed to stand in them for a minute or two; the glass of mineral water may likewise be warmed by the addition of a little hot water or hot milk. Sometimes the mineral water cannot be borne on an empty stomach, and in such cases the patient may drink a cup of tea or coffee first, or may mix milk or whey with the mineral water.

The baths are only prescribed for a small proportion of patients; for example, such as suffer from the uric acid diathesis, kidney troubles, or to strengthen the action of the vesical musculature in atonic conditions. The water of the baths is usually warmed to 77° to 99° F., and their stimulating action (sea salt or alkali is sometimes added) is increased by the bubbles of carbonic acid gas which move along the skin of the bather. The baths are usually taken in the forenoon, an hour or thereabouts before the mid-day meal. They are not prescribed when there is any tendency to hæmorrhage.

It is, however, the operative skill of the resident medical men that has given Wildungen the reputation that it possesses as 'a surgical spa' in diseases of the urinary organs, a reputation largely due to the labours of the late Dr. Stöcker. Vesical calculi are got rid of by lithotrity, strictures of the urethra are dilated or cut, and other surgical methods of treatment are employed. It is not, of course, maintained that mere drinking of the waters can cause solution of vesical calculi, or relaxation of a urethral stricture, though it may render the condition for operative interference more favourable.

The greater part of the patients are men, but women come for gravel and for various urinary troubles; sometimes also for an irritable condition of the bladder, not due to cystitis, but secondary to other pelvic troubles.

The Helenenquelle, owing to its alkalinity, is preferred to the Georg-Victorquelle in cases of much irritability of the bladder with highly acid urine; and owing to its being more easily borne by the stomach, it is preferred in most cases at the commencement of the course, especially if there be any tendency to constipation. On the other hand, the Georg-Victorquelle is more suitable when there is much vesical catarrh, with alkalinity of the urine, or when there is phosphaturia without mucus or muco-pus. When the patient is anæmic the Stahlquelle often forms a useful adjunct to the cure.

When, in addition to the urinary trouble, there is a tendency to bronchitis, dyspepsia, or hypochondriasis, these associated disturbances are likely to be remedied by the alkaline waters of the Helenenquelle, the wholesome diet provided, and the fresh mountain forest air. Owing, however, to the special reputation of Wildungen, it is less known as a health resort for such disorders, when not associated with urinary troubles, than some other places with waters belonging to the same class.

The principal season lasts from May 10 to September 25, but patients can be received in Wildungen at all other times of the year.

Access: In about 22 hours by Cologne and the branch line from Wabern. *Accommodation:* Good.

Doctors: Marc, Reinhold, Rörig, Schmitz, Severin, Winkhaus.

Contrexéville (France, Department of Vosges).—The village (altitude 1,150 feet) is a station on the railway, 13 hours by train from Paris. There are several springs of cold earthy water, the most famous of which is the 'Source Pavillon,' with 1·1 per mille sulphate of calcium, and ·44 per mille each of the bicarbonates of calcium and magnesium. The Contrexéville waters are employed for drinking in rather large amounts, producing diuresis, and having a slightly laxative effect.

The reputation of the spa is very great for affections of the urinary organs, which are 'washed out' by the treatment; for uric acid gravel, and oxaluria, and for cystitis. There are stories of vesical calculi undergoing spontaneous fracture in the bladder, and being passed in the urine, whilst the patients are under treatment, but this does occasionally, though rarely, occur elsewhere, without drinking medicinal waters. Contrexéville may be of use for some gouty conditions in weak subjects.

Baths and hydro-therapeutic treatment are employed as adjuvants to the internal use of the waters. Most people find the life at Contrexéville very pleasant. The season lasts from June 1 to October 20.

Access: About 18 to 20 hours from London, either *viâ* Calais or Boulogne, and Laon.

Accommodation: Good.

Doctors: Debout d'Estrées, Aymé, Graux, Thiéry, etc.

Lippspringe (Prussia, Province of Westphalia), 5½ miles from railway station of Paderborn, lies at an altitude of 450 feet, in a plain, to some extent protected by the Teutoburg Forest on the north. The weak earthy waters of the Arminiusquelle (temperature 70° F.), having a total mineralisation of 2·4 per mille, contain about ·7 per mille each of sulphate of calcium and sulphate of sodium, with smaller quantities of earthy carbonates, ·015 bicarbonate of iron, and a considerable quantity of free nitrogen and carbonic acid gases.

These waters are used for drinking, bathing, and for inhalation of nitrogen gas, but in recent years less for inhalation than formerly. Lippspringe is resorted to for chronic bronchitis, remains of pleuritic effusion, and chronic pulmonary tuberculosis; in the treatment of the latter affection its reputation is partly due to the writings of the late Dr. Rohden. (Season: May 15 to September 15.)

Inselbad, ¼ hour's distance from Paderborn, is an establishment for the treatment of asthma and chronic affections of the respiratory organs. The 'Ottilienquelle' (temperature 58° F.) is a weakly mineralised earthy spring, containing 40 per mille volumes of nitrogen, and some carbonic acid gas. There are likewise a chalybeate spring, used for drinking, and a sulphur spring.

Auerbach (altitude about 320 feet), a pleasant village in the Grand Duchy of Hesse, half an hour from Darmstadt, with beautiful beech woods in the neighbourhood, is a summer resort, and possesses weak earthy mineral waters, used for bathing.

Gran (Hungary) has thermal earthy springs (temperature 68° F.), and bathing arrangements. It possesses likewise a strong 'bitter water,' containing 45 per mille sulphate of magnesium.

Weissenburg (Switzerland, Canton of Bern).—The

two bath establishments are situated at an altitude of about 2,880 feet, 3½ hours' distant from the railway station of Thun, in a well-wooded sheltered dell above the Simmenthal. The new establishment is situated rather lower, and in a broader part of the valley than the old one.

The mineral water (temperature 79° F.) contains about 1 per mille sulphate of calcium[1] and a smaller amount of sulphate of magnesium; owing to the small total mineralisation (1·39 per mille) it may be classed either in the earthy or in the indifferent group. It exerts a diuretic action, and is said to cause constipation at first, but later on relaxation of the bowels. The dose commenced with is small, about an ounce, but is increased gradually until about a pint or more is taken in the day. If much constipation is caused a little sulphate of magnesium can be added to the minute amount which the spring water naturally contains.

Affections of the respiratory organs, including the early stages of pulmonary tuberculosis, form the chief class of cases treated at Weissenburg; the climate and general hygienic conditions playing a part in the results obtained. The season lasts from May 15 to September 30.

Accommodation: Good.

Doctors: Huguenin (from Zurich), Enderlin.

Similar to the water of Weissenburg is that of VALS in Canton Grisons (temperature 77° to 79° F.), but the total mineralisation of the latter is 2·04 per mile. The establishment lies at an elevation of about 4,100 feet in the Valserthal, 5 minutes from Vals-Platz, 14 miles from Ilanz, but the baths are little used.

[1] Phosphate of calcium has been noted in the Weissenburg water, but, according to Stierlin's analysis, the total amount present is only ·0004 per mille.

Saxon (Switzerland, Canton of Valais), a station on the railway from Lausanne to Brigue, lies in the valley of the Rhone at an elevation of 1,560 feet. Its weakly mineralised earthy waters have a total mineralisation of about one per mille, and contain minute quantities of the bromides and iodides of calcium and magnesium, but the iodides are said by Dénériaz to be occasionally or short periods altogether absent.

The climate is not bracing; the heat is often excessive; and the place is during some of the summer months infested by mosquitoes.

Bergün, a village in the Grisons, Switzerland, is a summer resort, situated on the Albula route at an altitude of about 4,500 feet. It possesses an earthy (calcium sulphate) spring, containing a little bicarbonate of iron.

Attisholz (Switzerland, Canton Solothurn) lies in a wooded valley at an altitude of about 1,670 feet, ¾ hour by omnibus from Solothurn. It possesses an earthy mineral spring (temperature 60° F.), and is visited by persons from the neighbourhood.

Bagnères-de-Bigorre (France, Hautes-Pyrénées).—This fashionable French spa possesses four classes of waters: indifferent thermal, cold earthy, sulphurous, and chalybeate.

The town is beautifully situated at an altitude of 1,890 feet, in the valley of the Adour. The springs, some of which were already known to the Romans, vary very much in their properties. The least mineralised may be classed as simple thermal waters, having a temperature of 90°–95° F., and may be used in the same cases as other simple thermal waters. The chalybeate springs are used in anæmic and debilitated patients, and some effect is claimed for a trace of arsenic in them. The cold sulphur waters of Labassère (7½ miles from the town) are employed internally in catarrhs of the respiratory system, etc., whilst the sulphur Pinac spring, in the town itself, is used for baths. The earthy springs differ considerably in the amounts of their solid ingredients, and may be used in various digestive and urinary complaints. This spa has a special reputation for uterine troubles. The season is from the middle of June to the middle of October. Bagnères-de-Bigorre is also a favourite climatic station.

Access: By railway *viâ* Bordeaux and Tarbes.

Accommodation: Good.

Doctors: Bagnell, Candelle, Collongues, Dejeanne, etc.

Capvern (France, Hautes-Pyrénées), a station on the railway from Toulouse to Bayonne, has an altitude of 1,300 feet, and possesses weakly mineralised earthy waters, with about 1 per mille sulphate of calcium, and a temperature of 70° to 76° F.

Siradan (France, Hautes-Pyrénées) has an altitude of 1,470 feet, and is situated about 12 miles from Bagnères-de-Luchon. It possesses cold earthy springs (1·3 per mille sulphate of calcium), and cold weak chalybeate waters.

Aulus (altitude 2,550 feet), a French village (Department of Ariège), in a deep valley at the north of the Pyrenees, possesses three tepid earthy sulphated springs (temperature 68° F.). These waters exercise a laxative and diuretic action, and are said to be useful in troublesome cases of tertiary syphilis.

Encausse (France, Department of Haute-Garonne) possesses tepid earthy waters (temperature 71·6° F.), containing about 2 per mille sulphate of calcium.

Cransac (France, Department of Aveyron) is a village with a station on the railway from Rodez to Capdenac; it lies (altitude 980 feet) at the foot of the still active volcano of Le Montet. Cransac possesses cold earthy waters which contain, in addition to sulphate of calcium and sulphate of magnesium, small quantities of the sulphates of potassium, aluminium, iron, and magnesium. The 'Source Basse Richard' (about 2 per mille sulphate of magnesium, and a total mineralisation of a little over 4 per mille) has a laxative action, and is used in cases of dyspepsia with chronic constipation, jaundice, etc. It has likewise a reputation in chronic malaria with enlargement of the spleen.

In the mountain sides are crevices, which are used as natural hot-air baths (temperature 90° to 118° F.) for chronic rheumatic cases; the air in them contains sulphurous vapours.

Vittel (altitude 1,100 feet) and **Martigny-les-Bains** (altitude 1,200 feet) are stations on the railway four miles respectively to the north and south of Contrexéville (Department of Vosges). They possess cold earthy springs, resembling those of the latter spa, and used for similar classes of affections. Their seasons are from about the third week in May to about the third week in September.

Doctors: Patézon, Bouloumié, and Marucheau (at Vittel).

Pougues (France, Department of Nièvre) lies on the right bank of the Loire, about eight miles from Nevers. It possesses cold alkaline earthy waters (1·7 per mille of bicarbonate of calcium in the Saint-Léger spring), which are used for dyspeptic troubles.

Saint-Amand (France, Department Nord).—The town (altitude 100 feet) lies on the Scarpe, and is a station on the railway between Lille and Valenciennes. Its weak earthy waters (·8 per mille sulphate of calcium) have a temperature of 70° F., and smell slightly of sulphuretted hydrogen gas. Saint-Amand is chiefly known for its mud baths, employed in chronic rheumatism, neuralgias, stiff joints resulting from injury, etc. The mud used for the baths is a peculiar soil permeated with the mineral water, and contains 1·4 per cent. carbonate of iron and a considerable amount of sulphuretted hydrogen gas. Patients remain from two to five hours in their compartments of this mud, and can read and write whilst immersed in it. The season is June 1 to September 30.

Lucca (Bagni di Lucca; Italy, Province of Lucca). The baths (altitude 500 feet) are situated at the foot of the Apennines, 15 miles from the town of Lucca. The springs have a temperature of 100°–120° F., and may be classed in the thermal earthy group. Most of them have a total mineralisation of about 3 per mille (chiefly sulphates of calcium and sodium). The chief season at Lucca is during June and September, when the place is much resorted to by the inhabitants of Florence. Many of the visitors come merely for amusement and change of air.

Chianciano (Central Italy, not far from Montepulciano) lies in the valley of Chiana, at an altitude of about 1,800 feet. It is reached from the railway station of Asciano by half an hour's drive, and possesses thermal earthy waters (temperature 100° F.), chiefly used for bathing. The total mineralisation is between 3 and 4 per mille (chiefly sulphate and carbonate of calcium). There are likewise gaseous chalybeate springs.

Besides the foregoing there are several other Italian earthy mineral springs, now comparatively little known, but some of them celebrated in ancient times.

Urberoaga de Alzola (Spain, a few hours' drive

from San Sebastian) is picturesquely situated in a gorge with beautiful environs. It possesses weak alkaline earthy waters (temperature about 87° F.), and has been somewhat misleadingly called the 'Spanish Vichy.' It has a reputation in affections of the bladder and urinary organs. The waters are used both internally and externally.

CHAPTER XVI

TABLE WATERS, AND OTHER VERY WEAKLY MINERALISED COLD WATERS

'Table waters' are feebly mineralised waters, containing more or less free carbonic acid gas, usually a large quantity, and may therefore likewise be termed 'simple gaseous' or 'simple acidulated waters' (in German, 'Einfache Säuerlinge'), or, when none of the gas has been artificially added, they may be called 'natural simple aërated waters.'

These waters may be of some use in medicine. They mostly contain minute quantities of bicarbonate of sodium, or of bicarbonate of calcium, or of both bicarbonates, and these may, in association with the carbonic acid gas, exercise a favourable effect in dyspeptic conditions. The carbonic acid gas stimulates the nerves and musculature of the stomach; in moderate quantities it aids digestion, promotes peristalsis, and relieves dyspeptic feelings; it probably also exerts some diuretic influence.

Such waters, however, are more frequently used as agreeable table drinks than for strictly medical purposes. Needless to say, iron salts in any considerable quantity and much bicarbonate of sodium mix badly with wines. The amount of solids contained in table waters should not be sufficient to give them any definite taste; and it is an advantage if there is much carbonic acid gas present (sometimes additional CO_2 is added before bottling), so as to prevent the precipitation of the mineral constituents and enable the water to 'keep well.' Much, however, of the temporary popular preference of particular 'table waters' over others depends on mere fashion and advertisement.

One of the great advantages which these waters have over many of the ordinary manufactured aërated waters is that the perfect purity of the water may be almost certainly relied on[1]—an inestimable advantage when there is reason to suspect that the ordinary drinking water of a town may be contaminated. The constant use of large quantities of highly gaseous table waters, whether natural or artificial, is, however, a habit not to be recommended.

Most of these waters are well known by advertisements, and as most of them have comparatively little to do with ordinary spa treatment, it will be sufficient here to enumerate them. They can be roughly divided into three classes, according as their mineral constituents show them to be weakly mineralised members of (1) the simple alkaline group of mineral waters, (2) the muriated alkaline group, or (3) the earthy group; in the latter case they contain small quantities of the carbonates of calcium or magnesium.

In the first group may be placed: APOLLINARIS (near Neuenahr), the JOHANNIS spring at Zollhaus, GEROLSTEIN, BIRRESBORN, TOENNISTEIN, all in Rhenish Prussia; OBERLAHNSTEIN, near Ems; TEINACH (the Hirschquelle), in Würtemberg; SULTZMATT, in Alsace; GIESSHUEBL

[1] The same can, however, be said for those artificial table waters in the manufacture of which distilled water only is used, or water which has been filtered through properly kept Pasteur-Chamberland, Berkefeld, or other reliable filters.

and KRONDORF, near Karlsbad, in Bohemia; SAINT-GALMIER, COUZAN (or SAIL-SOUS-COUZAN), RENAISON, CHÂTEAUNEUF, and SAINT-ALBAN, in France. Of these waters, Birresborn contains as much as 2·8 per mille bicarbonate of sodium, and is therefore rather strongly alkaline for an ordinary 'table water.' Bilin (3·3 per mille) and Fachingen (3·5 per mille) contain too much bicarbonate of sodium to be classed as 'table waters.'

In the second group may be placed ROISDORF, in Rhenish Prussia; ROSBACH, near Homburg, the KRONTHALBRUNNEN and the WILHELMSQUELLE at Kronthal, the TAUNUSQUELLE near Frankfurt, GEILNAU and SELTERS (or Niederselters), all in the Prussian Province of Hesse-Nassau; and the ACQUA ACETOSA, near Rome.

The third group includes the following: BELLTHAL, in Rhenish Prussia; the SELZERBRUNNEN, in Hesse-Darmstadt; the spring of GOEPPINGEN, in Würtemberg, mentioned by Paracelsus; and EVIAN, CONDILLAC, and CHÂTELDON, in France.

Many spas, described in other chapters, besides their better known, more active mineral springs, possess also weakly mineralised gaseous waters, which can or could be employed as simple acidulated waters. Amongst these are the Ludwigsbrunnen, at Nauheim; the Dorotheenquelle, at Karlsbad, in Bohemia; the Lindenquelle, at Schwalbach; the Wernarzerquelle, at Brückenau, etc.

Many waters used as table waters contain small amounts of iron, sometimes more than is advisable for ordinary table use; thus Saint-Alban water has over ·02 per mille of the bicarbonate of iron. Some table waters are sufficiently mineralised to be mentioned separately in other groups; thus, Birresborn and Toennistein are mentioned likewise amongst the simple alkaline waters. In some cases arrangements for the accommodation of visitors, with bath establishments, etc., exist at very weakly mineralised springs, notably at Evian and Giesshuebl, which we must therefore mention separately as spas.

Giesshuebl-Puchstein (Bohemia) is pleasantly situated in the valley of the Eger, on both banks of the stream, about six miles distant from Karlsbad. There is a bath establishment, but its mineral water is chiefly exported for use as an alkaline table water.

Evian-les-Bains (France, Savoy) is situated on the Lake of Geneva, opposite Lausanne, at an altitude of 1,240 feet. The cold alkaline waters are so feebly mineralised that they may almost be regarded as pure waters. Like simple water, the waters of the Evian springs exert a diuretic action; and they have a reputation in affections of the urinary organs and the uric acid diathesis. They are used both for drinking and bathing. In the case of some anæmic and cachectic patients the neighbouring chalybeate water of Amphion (see p. 193) is used simultaneously.

Doctors: Bordet, Taberlet, Chiais, etc.

Thonon (France, Haute-Savoie) has cold weakly mineralised waters, similar to those of Evian-les-Bains. The town lies on the southern shore of the Lake of Geneva, five miles to the west of Evian, but on a cliff about 130 feet above the lake.

With the springs of Evian and Thonon may be classed the cold weakly mineralised springs of Malvern and Ilkley, in England, and other nearly pure water springs much used in former times for their supposed special therapeutic effects, but in modern times, if used medicinally at all, used on ordinary hydro-therapeutic principles, sometimes in connection with special establishments. A great number of cold weakly mineralised springs, having a therapeutic reputation, exist in different parts of Europe. Some of them are classed as weakly mineralised earthy waters ; others contain so much free carbonic acid gas that they belong to the simple acidulated group ; others again, though they contain only minute quantities of the bicarbonate of iron, sometimes considerably below ·01 per mille, are yet classed as chalybeate waters.

One finds the cold waters of EMPFING (or WILDBAD EMPFING) and of ADELHOLZEN (or WILDBAD ADELHOLZEN) in Upper Bavaria, with a total mineralisation of under one-half per mille, still classed in the alkaline earthy group, and so also the cold waters of REHBURG in Hanover, with a total mineralisation of about 1 per mille.

There are also cold weakly mineralised springs with or without much free carbonic acid gas, containing a minute quantity of some special constituent, for which a particular therapeutic effect has been claimed, and these springs it is most convenient to place in the present group. Such are the springs of Saint-Christau, containing a minute quantity of sulphate of copper, and the 'iodine-springs' of Krankenheil. The gaseous weak compound chalybeate springs of Fideris in Switzerland, Fuered in Hungary, and Schwalheim near Nauheim (*q.v.*), contain only about ·01 per mille bicarbonate of iron, and may be fitly mentioned in this chapter.

Krankenheil-Tölz (Upper Bavaria).—Krankenheil is beautifully situated on the northern slope of the Blomberg, at an elevation of 1,130 feet above the sea. It is separated from Tölz by the Isar. Its cold weakly mineralised waters are most conveniently classed in this group. They contain ·19 to ·33 per mille bicarbonate of sodium, ·03 to ·29 per mille chloride of sodium, about

·001 per mille iodide of sodium, and a little sulphuretted hydrogen gas.

With such a weak mineralisation it is difficult to see what special therapeutic effect the water can have, but Krankenheil has a reputation in scrofulous affections, chronic endometritis, skin eruptions, etc. The season lasts from May 15 to October 1. It may be noted that salts derived from the Krankenheil waters, soaps made with the salts, and the concentrated mineral water, are all made use of in the treatment of patients.

Zaizon (Transylvania, altitude 2,590 feet), a spa visited chiefly by women and children, contains the weak gaseous muriated-alkaline 'Ferdinand's spring' (1·3 bicarbonate of sodium, ·6 common salt) which attracted some attention on account of its being said (probably by error) to contain ·25 per mille iodide of sodium. There are also weak chalybeate waters.

Saint-Christau (France, Department of Basses-Pyrénées), a small spa situated at an altitude of 985 feet, in the narrow Pyrenean Valley of Aspe, possesses feebly mineralised cold earthy waters (total solids about ·3 per mille), in which are minute quantities of the sulphates of iron (·004 per mille) and copper (·00035 per mille). The waters, besides being used for drinking and bathing, have been employed in a finely pulverised form for chronic laryngitis and pharyngitis, and for the eye in chronic blepharitis and conjunctivitis.

Fuered (Balaton-Füred), a popular spa in Hungary, at an altitude of 480 feet, is beautifully situated on the Plattensee, one hour by steamer from the railway station of Sio-Fok. Its weakly mineralised waters might be classed in the sulphated alkaline, in the earthy, or in the chalybeate group, but, considering their probable mode of action, are most conveniently classed in the present group, to follow the simple acidulated and other weakly mineralised waters. The favourite well used for drinking is the gaseous 'Franz-Josephs-Quelle' in the Kurplatz, which contains about ·8 per mille each of carbonate of calcium and sulphate of sodium, ·11 per mille of carbonate of sodium, ·01 per mille of bicarbonate of iron, and 1,207 per mille volumes of carbonic acid gas. The weakly mineralised waters of the lake (containing 54 volumes per mille of carbonic acid gas) and the mud from its banks are both used for baths. The season is from May 15 to September 15.

Fideris (Switzerland, see p. 191) must likewise be mentioned in this chapter. It lies at an elevation of 3,460 feet in the Praettigau valley, and its cold gaseous waters contain only a minute quantity of iron (·01 per mille of the bicarbonate) and a total of under 2 per mille solids.

CHAPTER XVII

MARINE SPAS

THE foregoing chapters have been devoted to inland spas; marine spas are so generally appreciated, and so much has been written about them in England, that only a few words need now be devoted to them. The sea is really a mineral water, and sea baths act in the same way as the inland fairly strong salt or brine baths, called by the Germans 'Soolbäder' (see p. 78). There are, however, important differences between the 'Soolbäder' and sea baths. In sea bathing, or, as it usually is, 'surf bathing,' there is the charm and freshness of bathing on the open sea shore, and there is the mechanical stimulation on the skin by the impact of the waves and movement of the water. This is absent in the Soolbäder (unless, indeed, the waves be artificially imitated); they rather resemble the taking of sea-water baths in closed establishments or at home, which can also be done at marine spas, if preferred, in the case of very delicate and timid people. In the latter case the sea water may be artificially heated, and thus made to resemble the warm Soolbäder.

Sea bathing

Internal use of sea water

Although the internal administration of small doses of diluted sea water has been occasionally advocated and sometimes carried out with apparent good results, the unpleasant taste is not likely to render the custom at all general or 'fashionable.'

Practically, therefore, sea water is only used in the form of baths. Nor does it make a very great difference which sea water is used, though the amount of solids contained in sea water from the Baltic is less by half, and that from the Mediterranean is slightly greater than that from the German Ocean or Atlantic, which contains about 3 per cent. of common salt.

Difference in sea waters

Much of the effect of the seaside depends on the 'freshness' of the air, owing to the constant breezes. During the daytime the surface of the land gets heated more than that of the sea, the hot land heats the lower layer of air, which becomes lighter and rises, its place being taken by the cooler air from the sea; hence the prevalence of sea breezes during the heat of the day. After sunset the surface of the land cools down more rapidly than the surface of the sea; hence the prevalence of land breezes in the evening. It is this perpetual movement in the air that makes the seaside so enjoyable during the heat of summer, and gives it a certain bracing effect which is useful to those who are debilitated from overwork and to convalescents. There are some people who become 'bilious' and constipated from the effect of sea air, especially at the more bracing places, and for whom simple country air or mountain air is likely to be more useful. In many cases, however, this inconvenience may be avoided by diminution of food, and increase of excretions through the influence of aperient waters or drugs.

Sea air

Sea bathing and sea air are especially serviceable in the treatment of scrofulous affections in children; even tuberculous affections, in which surgical interference is necessary, perhaps do better when operated on at the seaside than in the hospital of a large town. In many

treatment

different classes of debilitated or anæmic patients the seaside is likely to be of use with or without special treatment. In all cases much attention must be given to the separate tendencies of the individual patients, as well as to the affection from which they suffer or have been suffering. As a general rule the more bracing localities are more suited for those individuals who retain tolerably good power of reaction to cold, whereas the milder climates are suited to those with little power of reaction and to patients of very irritable nervous temperaments. Sea bathing may aggravate an eczematous eruption or bring out an urticarial or other rash, or in some cases may be followed by headache or too great a feeling of fatigue; in such cases it should be abstained from temporarily or permanently, or else baths of very short duration (always to be recommended at the commencement of a course) should be tried; or possibly baths of warmed sea-water taken in the house may be found suitable to begin with.

It is doubtless owing to the improvement of the general health, by sea air and sea bathing, that some marine health resorts have acquired a reputation in cases of impotence. It is thus also that sea bathing may cure leucorrhœa and amenorrhœa, associated with anæmia; in fact, many different affections, when partially or wholly due to a depressed state of the general health, may be cured or relieved by seaside treatment.

Selection of a marine spa

There are questions of varying importance to be considered in the suitability of marine spas, such as the character of the shore and its suitability for bathing, the position of the place, its surroundings, and its climate at different seasons of the year; and, lastly, its accommodation, hygienic arrangements, and the amusements afforded to visitors.

Each place has its advantages for different invalids, and in some cases its disadvantages. Sometimes the shore is precipitous, or there is not sufficient beach for bathing, sometimes the sandy shore is so level and extensive that persons bathing have to go a considerable distance out, in order to reach a sufficient depth of water to cover the body—a circumstance not without its advantages in the case of children. Sometimes there is a considerable descent from the houses to the shore, so that unless special arrangements exist, invalids may find the getting up and down hill between their houses and the sea tiring. The accommodation may at times be hardly sufficient for the sudden influx of visitors. Whilst some spas are too crowded, others are described as dull and without amusements. There may be disagreeable smells, or the drainage arrangements may be defective, but this is less likely to be the case in England than in other countries.

England is notorious for the trouble bestowed on drainage, and the freedom of its towns from disagreeable smells; but it must be owned that at such a popular seaside health resort as Margate the smell which comes from the harbour during low tide, on a hot day, is anything but agreeable, though it need not necessarily be due to defective drainage arrangements.

The main points to be considered in selecting a marine spa are the climate, the hygienic arrangements, and the surroundings. The time of year must likewise be considered; many of the fresh bracing seaside places of great use in summer are too cold for invalids in winter. If, therefore, a patient applies for advice in winter, a milder health resort has to be recommended, though in summer and autumn a more bracing locality might have to be suggested to the same person.

The most convenient classification of marine spas is that by the main characteristics of their climate; in the following pages they have been roughly arranged into different groups, according as their climate may be termed dry, moist, or of medium humidity, and as it is warm or cold.

Warm dry marine places

Amongst the *warm dry marine* health resorts the most important and well known are those of the Western Riviera, including St. Raphael with Valescure, Hyères with Costebelle, Cannes with Cannet and Grasse (the latter at an altitude of about 1,000 feet, nine miles inland from Cannes), Antibes, Nice (with Cimiez), Villafranca, Beaulieu, Eze, Monte Carlo, Cap Martin, Mentone, Bordighera, San Remo, and Alassio.[1] In spite of the frequency of winds,

[1] *Costebelle*, which consists only of hotels, is beautifully situated on the south slope of a hill covered with pine and 'maquis,' and tolerably protected from the north-west, but less so from the north-east; the latter, however, prevails not so frequently, and acts on the whole less injuriously at this coast than the 'mistral' or north-west wind. *Hyères* is not so well sheltered, though it has increased in size, and possesses much improved hotel accommodation. *St. Raphael* is much less protected from the winds, and *Valescure* has not developed as it was expected to do. Many boulevards and villas which had been commenced at Valescure, have remained uncompleted; the pine trees are too much scattered to afford protection, and although the Estérel Hills afford a slight protection from the north-east, the place lies fully exposed to the 'mistral.'

In the last twenty years *Cannes* has grown enormously as far as hotels and villas are concerned, particularly in the eastern quarter, where the houses stretch out for two or three miles amongst the pine trees on the slopes of the hills towards Antibes. The water supply and the drainage are much improved. *Grasse*, whose beautiful position somewhat reminds one of Les Avants near Montreux, is sheltered by considerable heights from the west, north-west, north, and north-east; the air is fresh, the views and walks are delightful, and the accommodation at the Grand Hotel in the highest part of the town, 1,100 feet above sea level, is good. Many cases of neuralgia and asthma find relief from these complaints, which had been aggravated at the seashore. [In the hills, four and a half hours to the north-west of Grasse,

clouds of dust, sudden changes of temperature, and evening fogs, those who require more sunlight, warmth, and dry-

is *Thorenc*, nearly 4,000 feet above sea-level, a summer health resort, near to a pine forest, which is in process of development, and may turn out of great value to invalids at the Riviera.]

Nice, like Cannes, has enormously increased in size during the past twenty years, and the appearance and hygienic arrangements are likewise improved. *Cimiez* especially has been further developed, and is found by most nervous people less exciting. *Beaulieu*, on the railway between Nice and Monte Carlo, is a small strip of land between the sea and high rocks, which shelter it from the N.N.E., and partly from the north-west. With the sole exception of the eastern bay of Mentone, Beaulieu and Eze are the most sheltered spots on the Riviera, and the irradiation by the sun's rays reflected from the rocks has procured for Beaulieu the name of 'Petite Afrique.' The very limited strip of ground around the railway station of *Eze* (Italian Eza), the next station on the way to Monte Carlo, is, if anything, still more sheltered than Beaulieu; behind it, the old robber's stronghold of Eze forms a most picturesque object, crowning the steep rocks, about 1,300 feet above sea level.

Cap Martin, between Monte Carlo and Mentone, possesses now one of the best situated and best arranged hotels on the Riviera. It lies at an elevation of about 150 feet above the Mediterranean, and is surrounded by a large pine forest with an undergrowth of Rosemary, Myrtle, Lentiscus, and Cistus. It has the great advantage of being as good as free from dust, and exercises a more soothing influence on the nervous system than does either Mentone or Monte Carlo. The pine forest with its undergrowth gives shelter from wind and sun, and fragrancy to the air. At present it cannot be regarded as a good health resort for serious pulmonary cases, on account of the large element of mere pleasure-seekers amongst the visitors; but there are good sites in the fir plantations for other establishments, which may be turned to greater advantage for invalids.

Bordighera has grown much in the last twenty years, but has retained its old character of a quiet health resort. The hotels and villas which lie away from the sea, in the olive plantations, have more shelter and less dust. The villas of the neighbouring *Borghetto* (usually included under the name Bordighera) are those best protected from the wind. The air of Bordighera is on the whole fresher than that of Mentone and San Remo.

In *San Remo* many new hotels and villas have made their appearance, as well in the eastern as in the western portions of the town, partly perhaps since the late Emperor Frederick made a trial of the climate.

ness of air than they can get in their own country, often gain appetite, and become healthier in mind and body, by residence at one of these places. The best time of year for these localities is generally from the end of October to the end of April. Amongst invalids likely to be benefited on the Western Riviera are: delicate and scrofulous patients with low power of resistance; some of those affected with chronic or quiescent pulmonary tuberculosis, or with catarrh of the respiratory or intestinal mucous membrane, or suffering from the remains of pulmonary affections; gouty and rheumatic patients extremely sensitive to cold and damp; lastly, those whose power of resistance is temporarily or permanently very much lowered by previous disease, injuries, or by premature senility. As in all cases of disease, so here it is not the nature of the disease which has exclusively to be considered, but the individual peculiarities and tendencies of the patient. Those subjects of pulmonary tuberculosis who are harassed by a dry nervous cough, laryngeal irritability, or in whom every slight cold produces a febrile temperature, generally do better in moister and more equable climates, such as Ajaccio, Algiers, and Arcachon. For hysterical patients and those suffering from neuralgic conditions, moister, cooler, and more elevated regions are mostly preferable.

Sorrento and Castellamare, in the beautiful Bay of Naples, though probably too hot for most natives of Northern Europe in the height of summer, would be delightful in the spring and autumn. Sea bathing can be

A mild summer health resort is going to be opened at *Ormea* (2,460 feet above sea level), about five hours' drive by a good carriage road from Oneglia, a railway station between San Remo and Alassio. *Alassio* has not much increased in size, though for many invalids it is preferable to San Remo. The old town proper lies quite close to the sea, but the surrounding semicircle of hills, especially the slopes facing the south and south-west, afford good sites, which are warmer in winter and cooler in summer than the level ground on which the old town stands.

had at both places, and at Castellamare there are likewise the earthy muriated mineral springs (see p. 126). The islands of Ischia and Capri are rather too much exposed for winter residence. Naples itself, and the almost equally beautiful and famous Salerno, are unfortunately still open to suspicion as to hygienic arrangements, but Naples is already much improved; the malarious air from the marshes near Pæstum at times reaches Salerno. Amalfi, on the northern shore of the Bay of Salerno, about twelve miles from the city of Salerno, has an exhilarating and healthy situation. Though very sunny in winter, it is only partly sheltered from the north.

To this class of health resorts belong also several on the Mediterranean coast of Spain. Amongst these may be mentioned Barcelona, and the warmer Alicante, and Malaga, described by Francis as the mildest place in Europe. The latter has a dry sandy soil, a south-eastern aspect, and is protected by a semicircle of mountains from the north and north-east winds, but is exposed to the biting dry north-west wind. Valencia belongs rather to the class of warm moderately humid climates.

The *cold and dry* marine localities are not yet used as health resorts. No European places can well be classed like Madeira amongst the *warm humid* marine health resorts, whilst the *cold humid* marine climates of the Hebrides, Orkney and Shetland islands, though very interesting for their great equability of temperature, are rarely used in the treatment of disease. There remain therefore only the marine spas of *medium humidity* to be mentioned, but amongst these are included by far the greater number of European marine health resorts.

Amongst the *warmer* localities of medium humidity some of the least humid ones are the towns of the 'Eastern Riviera:' Viareggio, Spezia, Chiavari, Rapallo,

The warmer marine places of medium humidity

Santa Margherita, and Nervi. These places, excepting Nervi,[1] are somewhat less sheltered from the cold winds, and have a relatively higher humidity than the localities previously mentioned in the 'Western Riviera.' Pisa now lies a few miles inland, and hence its climate is not strictly marine. Genoa is windy and rainy, but Pegli, about six miles west of Genoa, has a more equable and sheltered climate, though it has a greater relative humidity than the localities further west, that is, in the Western Riviera proper. Patients who have wintered in the Western Riviera may stay at Pegli in spring on their way to the Swiss or Italian lakes.

Venice is not so warm as either of the Rivieras, and is not sheltered from the cold north wind. It does not quite deserve its former reputation in phthisis, but its freedom from dust is a great advantage, and cases of arrested phthisis with a tendency to irritable cough, and some cases of nervous irritability, may be recommended to Venice, especially in March and April, when other places have also great defects.[2]

Amongst the Ionian Islands the towns of Corfu and Zante, though very hot in summer, have a too uncertain and variable climate in winter.

[1] Nervi is the only really sheltered spot in the Eastern Riviera, and offers as great natural advantages as almost any locality of the Western Riviera. Most of the place is private property of rich Italian noble families, but the Eden Hotel, well situated on the slope, affords good accommodation to the visitors (who are chiefly Germans). One of the principal features of Nervi is the walk along the picturesque rocky coast, well sheltered and entirely free from dust. It is perhaps the finest walk at any marine spa in Europe, and it is due in a great degree, we understand, to the exertions of Dr. Schetelig.

[2] Rheumatism is very prevalent in Venice, and it is absolutely necessary for invalids to avoid rooms near the ground-floor or deprived of direct sunlight. The *Lido* island, which may be regarded as part of Venice, has good arrangements for sea bathing, and would be an excellent marine health resort if it were quite free from malaria.

In Sicily the famous towns of Syracuse, Palermo, and Catania must be mentioned; though very hot and sunny, they are too exposed to disagreeable winds for many serious invalids. Acireale (see p. 244), on the railway between Catania and Messina, has to invalids the advantage of being a smaller town than Catania.

Ajaccio,[1] in the island of Corsica, faces the south-west, is sheltered from cold winds, and is favoured by its freedom from dust and mosquitoes; there are beautiful walks and drives to be enjoyed in the neighbourhood, the roads are excellent for excursions, the accommodation is good, and the people are friendly.

Valencia belongs to the medium humid localities, and not to the dry ones, like some other places on the east coast of Spain previously mentioned. Its mild and equable climate is to some extent spoiled by the irrigations of adjacent rice-fields.

The climates of Lisbon and other towns, on the western coast of Spain and Portugal, are too changeable to render them suitable spots for invalids to stay at.

Biarritz and St. Jean-de-Luz in the south-west of France, on the coast of the Bay of Biscay, are exposed to the prevailing winds, and are bracing to most persons. Though there is much rain, the air rarely seems damp, owing to the dry soil rapidly absorbing the rain. They are pleasant autumn and spring resorts for bathing, and are to be recommended in cachectic conditions from long residence in hot climates, in patients without organic disease, and in certain hypochondriacal conditions.

[1] In regard to Ajaccio the situation on granite, the absence of dust, the shelter from wind, are specially to be noted, as well as a peculiar aromatic condition of the air, due to the dense 'maquis' covering all the surrounding hills. The 'maquis' or Corsican 'bush' is composed chiefly of arbutus, cistus, lentiscus, myrtle and heath. One can well imagine how Napoleon, when at St. Helena, said that he would recognise Corsica with shut eyes, by the aroma of the air.

Arcachon, further north, about nine miles from the actual coast, lies in a pine forest, near a large basin of salt water, connected by a narrow channel with the sea. According to Dr. Burney Yeo its climate is 'mild and soothing, and is especially suitable to cases of irritable bronchial or laryngeal catarrh,' and 'to cases of phthisis with tendency to congestion or inflammatory complications.'

The cooler marine places of medium humidity

Amongst the *cooler* localities of medium humidity must be included the numerous marine spas of Great Britain and Ireland, and those on the north-west and north coasts of France, and on the coasts of Belgium, Holland, and Germany.

The climate of Great Britain and Ireland is made warmer than other countries of the same latitude by the Gulf Stream and by moist winds warmed by warm currents in the Atlantic Ocean. The rainfall, though not much greater, is more equally distributed over the different seasons than it is in more southern countries. The clouds so common in the British sky, whilst they to some extent keep off the warmth and light of the sun during the day, check the loss of heat by radiation during the night, and so tend to equalise the night and day temperatures, and prevent the chilliness so often felt at sunset in the warmer and brighter Riviera. The hygienic conditions and accommodation at English seaside places are, moreover, usually very good, an advantage that they possess over many foreign places.

There is a considerable difference between the climate of the west and south-west coast, and that of the east and south-east coast of England; the latter are colder and drier than the former, whilst the main part of the south coast combines the dryness of the latter with the warmth of the former. It is in winter that the difference in temperature shows itself chiefly; hence some of the warmer seaside spas may be chosen as winter health resorts. Some of these will be mentioned first.

Queenstown, in Cork Harbour, Ireland, is well sheltered from the north, and is as warm as Torquay in Devonshire. Glengariff, in Bantry Bay, has a similar climate. Rothesay, in the island of Bute, is amongst the places on the west coast of Scotland which enjoy a comparatively mild winter climate.

The Scilly Islands have been said by Dr. Tripe to possess 'the most equable winter temperature in the British Islands, if not in all Europe.' Penzance and Falmouth, in Cornwall, have a very equable climate, though not so well sheltered from the winds as Torquay in Devonshire; Falmouth has recently (1896) been alluded to by Sir Joseph Fayrer, who has personal experience of its climate, in terms of high commendation. Torquay is said to be drier than other places of South Devon. The parts of Torquay, further up the hills, away from the sea, are less relaxing than the part nearer the sea.

Teignmouth has not such an equable climate as Torquay, and is not sufficiently sheltered for winter residence. Dawlish is more suited as a winter residence for invalids, but owing to east winds is less serviceable during the spring. The new portion of Exmouth is fairly sheltered, but subject to occasional river fogs.

Sidmouth is almost as well sheltered as Torquay, and has great advantages as a winter health resort. The new bath establishment at Sidmouth offers facilities for warm sea-water baths, etc.; the 'Aix douche-massage,' and the 'Nauheim treatment' of heart affections are likewise said to have been introduced there.

Salcombe, owing to its protected position, is one of

the warmest spots in England, but the walks for invalids are too limited. Many other places on the south-west coast might be mentioned in this group, as well as Lynmouth, Lynton, Ilfracombe, and some other towns on the north coast of Devonshire and Cornwall.

Two of the most important winter health resorts on the coast of England are Bournemouth in Hampshire, and the Undercliff of the Isle of Wight. Bournemouth, on account of its plantations of pine trees, has been compared with Arcachon in France. There is more wind at Bournemouth than at Torquay, but the place is tolerably sheltered from the north, north-east, and, to some extent, from the east winds, and the air is less relaxing than at Torquay. The sand and sandstone on which the town is built, absorb the rain and help to keep the atmosphere fairly dry. The popularity of Bournemouth as a health resort and as a wintering place for patients with pulmonary and bronchitic troubles is witnessed to by the enormous extension of the town along the coast in recent years. The neighbouring Branksome is practically a continuation of Bournemouth.

The Undercliff of the Isle of Wight is a kind of terrace extending for about six miles in length, from near Bonchurch to Blackgang Chine. The warmth of the sun is increased by reflection from the cliffs and the sea. The soil is of chalk and sandstone, absorbing water, and so leaving the surface dry. Its position is sheltered from the north, north-east, north-west, west, and partly from the south-west. The scenery is beautiful, and the climate is mild and equable, yet fairly dry, and not relaxing. It is often suitable in early stages of phthisis, and in chronic catarrhal conditions of the respiratory system, in some scrofulous, anæmic, and debilitated conditions, and in slow convalescence from acute diseases. The great fame of this portion of the Isle of Wight has been justified by the satisfactory results of the National Hospital for Consumption at Ventnor.

Pwllheli, in Cardigan Bay, has some claim to be considered a winter resort, owing to the shelter afforded by the mountains. Hastings and St. Leonards in Sussex, and Llandudno in North Wales, are best known as summer resorts, but might be used as winter resorts by those able to bear a certain amount of cold wind.

The places in the British Islands suitable for seaside residence in summer are too numerous to be all mentioned. Their winter climate is colder than those previously mentioned; in their summer climate they differ less. In Scotland: Nairn, North Berwick, St. Andrews, and Portobello may be mentioned. Beginning from the north in Yorkshire, we have Redcar, Saltburn, Whitby, Scarborough, Filey, and Bridlington. In Norfolk and Suffolk: Hunstanton, Wells, Cromer, Yarmouth, Lowestoft, Aldborough, and Felixstowe. In Essex: Walton, Clacton, Southend. In Kent: Herne Bay, Birchington, Westgate, Margate (with its well-known infirmary for scrofula), Broadstairs, Ramsgate (with the adjoining St. Lawrence), Deal, Walmer, St. Margaret's Bay, Dover, Folkestone, Sandgate, Hythe. The towns mentioned after Dover are to some extent sheltered from the north, and are warmer than the east coast localities.

We now proceed westwards along the south coast in our enumeration, passing Hastings, with St. Leonards, Eastbourne, Seaford, Brighton,[1] Worthing, Littlehampton,

[1] Brighton has not been mentioned amongst marine stations for winter on account of its want of shelter from the east, which, in the

Bognor, and come to the various seaside towns of the Isle of Wight not included in the Undercliff, *i.e.* Shanklin, Sandown, Sea View, Ryde, Cowes, Yarmouth, Alum Bay, and Freshwater. Further west along the coast are Southsea, Swanage, Weymouth, Lyme Regis, and places previously mentioned as variously suitable for winter residence. In many of these much depends on the position of the house in the town, different portions of the same town being more sheltered from cold winds than others, and the sea air being most felt in the parts nearest the sea. The sea air is more felt in the Channel Islands than in any of these places.

Along the north coast of Cornwall, Devon, and Somerset, we have New Quay, Bude, Barnstaple, Ilfracombe, Lynton, Lynmouth, Minehead, Weston-super-Mare, and Clevedon. The three latter, on the Bristol Channel, have the disadvantage of large muddy sand-fields during low water, but have beautiful walks: their climate is less fresh than the north coast of Cornwall.

On the coast of Wales there are Tenby, Aberystwith, Barmouth, Pwllheli (already mentioned), Beaumaris (in the Isle of Anglesea), Llandudno (already mentioned), and other places.

Further north than Wales one comes to New Brighton, Southport, Blackpool, Fleetwood, and Grange in Morecambe Bay, the latter in a beautiful and sheltered position. Silloth, in Cumberland, on the Solway Firth,

case of invalids, renders special precautions necessary during the months of February, March, and April. The north wind may likewise be very unpleasantly felt, except in the so-called 'Madeira Walks.' From late autumn to January, Brighton may be a good station for invalids, but it is not likely to become a real winter health resort, unless some sort of large winter garden or 'glass-palace' be erected, with complete protection from the east, north, and north-west, so that invalids can daily spend four to six hours in it, and take their exercise there.

has a mild and comparatively dry climate. Douglas and Ramsey, in the Isle of Man, have naturally a completely marine climate, being situated in the midst of the Irish Sea.

On the west coast of Scotland there are Ardrossan, in Ayrshire; Dunoon, Largs, and other places near the Firth of Clyde; and Rothesay, on the Island of Bute (previously mentioned), all of them very useful resorts for the industrial centres of the west.

In Ireland Bray, Howth, Kingstown, Dundrum, and Holywood on the eastern coast possess a mild and humid climate. Port Rush, near the Giant's Causeway, and Port Stewart, on the northern coast, are more bracing and less humid. On the east coast Bundoran in Donegal Bay, and Kilkee and Kilrush in Clare, are exposed to the influence of the Atlantic. On the south coast, Queenstown and Passage in Cork Harbour, Glengariff (already mentioned), and other localities have a mild, equable, humid climate.

We now come to foreign marine spas belonging to the cooler, moderately humid group. The climate of the north-west coast of France, especially Finisterre, somewhat resembles that of the south-west coast of England, but the north coast of France is drier and more bracing. It is on this coast that many popular summer health resorts are situated, their season being from July to September. Amongst them, beginning from the west, are Cherbourg and Dinard; the latter is popular with English and American families, as is also the neighbouring interesting old town of Dinan, which does not lie quite on the coast. Then come the simple and unpretending Cabourg, Beuzeval, and Villars-sur-Mer, and the more fashionable and expensive Trouville, with Deauville. Further east are Etretat—converted from a small fishing village into a seaside spa by the patronage of French artists—Fécamp and Dieppe; then St. Valéry-en-Caux, Tréport, Boulogne, and Calais—so well known to the English—and Dunkirk.

On the Belgian coast is Ostend, with its bracing air, unrivalled sands, and fine Cursaal on the 'Digue;' further east are the more recently constituted health resort of Blankenberghe, and the less pretentious Heyst. Scheveningen, two miles from the Hague, on the coast of Holland, is one of the best-famed seaside summer resorts on the Continent.

The German North Sea coast possesses many bracing seaside places, amongst which are the small islands of Borkum, Norderney, Baltrum, Langeoog, Wangeroog, etc., most of them probably little known to English and Americans. The Island of Heligoland, now belonging to Germany, has a thoroughly bracing marine climate and good bathing, and is much visited by North Germans. Further north are the Schleswig islands of Foehr and Sylt, with the marine spas of Wyk and Westerland respectively; Sylt possesses likewise a chalybeate spring.

The Baltic spas have the advantage of beautiful forests in their neighbourhood, but are less bracing than the North Sea health resorts. Amongst them may be mentioned Marienlyst, Düsternbrook, Travemünde, Doberan or Heiligen Damm (with a weak chalybeate spring); Sassnitz, Putbus and Binz on the island of Ruegen; Heringsdorf, Swinemünde, Misdroy, Dievenow, Kolberg (or Colberg), Ruegenwalde, and Cranz. Many bracing seaside places in Denmark, Norway, and Sweden, might likewise be mentioned.

CHAPTER XVIII

BALNEO-THERAPEUTIC MANAGEMENT IN DIFFERENT DISEASES AND MORBID CONDITIONS

In discussing the treatment of diseases or morbid conditions by spas, or by courses of the internal and external use of mineral waters, the physician and the patient must above all things drop the superstitious belief that the balneo-therapeutic treatment is something entirely different from the ordinary methods of treatment. The same considerations which guide the physician in the ordinary management of his patient in the deviations from the natural healthy conditions of life, must guide him in the use of mineral waters; but the means which he uses in prescribing spa treatment are much more complicated than those of the ordinary home treatment. They require an intimate knowledge of all the elements which come into play, when he resorts to spa treatment, and of the influences which they are likely to exercise on his patient. He has to regard not only the mineral waters which we can compare with the pharmaceutical elements, and which in themselves are often very complicated, but also the removal from home, the journey to, and the climate at the spa, the altered diet and accommodation and other hygienic elements, and above all the qualities of the physician who is to guide the patient. Besides, we must bear in mind that the physical influences of these agencies are in many cases aided, in some also counteracted, by the subtle influence of the mind. This influence is mostly powerful, especially in persons who are said to have 'nerves,' but it is by no means always calculable, and thus creates an element of uncertainty which does not exist to the same degree in prescribing pharmaceutical remedies or change of air at home.

Very often it happens that the diseased condition of the patient is complicated, and that the physician who advises spa treatment has to consider carefully which part of the diseased organism he is to act upon. He must form an idea of the nature and power of the constitution, and calculate in how far the different organs and systems can assist his attempt to restore the healthy working of the diseased organ or organs. We will endeavour to show this by an instance which is by no means rare, and occurs in numerous variations with more or less grave complications.

We will suppose that a person past the middle age suffers from frequent catarrhal affections of the bronchial tubes at the lower parts of the lungs, with imperfect contractions of the (probably dilated) heart; in consequence there is passive congestion of the liver, and possibly already of the kidneys, with a loaded urine, containing large amounts of urates and often small quantities of albumen. In such cases the physician must consider whether the patient's abdominal system will allow him the use of purging waters in order to relieve the portal system and the liver, and, through this, ease the action of the heart, and, by the more regular and powerful pumping action of the latter, the lungs; or whether, owing to the condition of the patient, he can only make very gentle demands on the bowels, and has to direct his action to the skin, and through this on the heart; in the latter case he may proceed by the use of warm baths, simple or saline or gaseous saline, in combination with, or without, carefully arranged exercises, or he may try restricted diet with digitalis or mercury, or both combined, or other pharmaceutical agents.

Often it occurs that the invalid has an affection which cannot be treated directly, it may be of the heart, of the kidneys, of the spleen, of the skin, or of the nerves, and that the medical adviser must recommend spas or climates by which the general health is improved, and the diseased part of the organism is indirectly drawn into the general improvement. On advising about spa treatment the physician must consider, as far as it lies in his power, all the influences which are likely to act on the patient on the way to the spa, during his stay at the latter, and during the first weeks or months afterwards. In the preceding portions of the book special chapters are devoted to the discussion of these influences, and also to the importance of the qualities and duties of the local physician, on whom, to a great degree, the success of the spa treatment depends, and to whose guidance, therefore, the invalid ought in all cases to entrust himself.

Although it has already been mentioned that the patient should not make the sometimes necessarily long journey to the health resort without rests, we consider it necessary to repeat once more that every precaution ought to be taken to render the travelling as little fatiguing as possible, so that the patient shall not arrive in an exhausted condition. For this purpose the route and the places where the journey is to be broken, must be carefully planned beforehand, as well as the time of the day for travelling; railway journeys in the hottest time of the day are often very injurious to delicate people.

With the description of the principal spas their suitability in the treatment of various affections is given; we will restrict ourselves therefore, in this chapter, to summarising what spas and spa treatment are applicable to the cure of different diseases.

The grouping of morbid affections cannot be quite strict, because different systems and organs are often affected in the same person, and general affections are mostly combined with local deviations; but for the sake of convenience we will at first consider some general or constitutional affections, and afterwards those of different systems and organs. For the description of the spas and the classes of spas which we suggest, we refer to the former part of the book.

Tardy convalescence

1. *Tardy convalescence* from acute diseases is an important subject, and requires most careful management. There is a certain degree of exhaustion of vitality; the blood and all the tissues are, so to say, watery, and the proportion of solid constituents is diminished. All the functions are without energy. The circulation is weak, the heart is irritable; a slight exertion may raise the pulse from 60 or 70 to 140, and more in the minute. The skin is often damp, and a slight wind or change of temperature may cause a chill and serious general disturbance. In many cases of this class spa treatment is less useful than other slightly stimulating influences, such as sea air in some, and forest or mountain air in other cases. It is often necessary in these conditions to assist the removal of the products of retrogressive metabolism, and we must consider whether this ought to be done by pharmaceutical remedies or by spas. In

the latter case generally the indifferent thermal spas (Chapter VI.) are useful, or the thermal 'Soolbäder,' such as Nauheim and Oeynhausen. For internal use the muriated waters (Chapter VII.) are preferable to the sulphated or alkaline waters. It frequently happens that the anæmia resulting from the acute disease requires chalybeate remedies, either pharmaceutical or in the form of mineral waters. Under all circumstances fatigue, great heat, and great cold, and the exposure to violent winds, must be avoided. At the same time the diet ought to be carefully superintended.

General debility

2. *General debility* is a term which may not be scientific, but the condition is real and demands sympathy and help. The symptoms are in many persons similar to those of tardy convalescence, although no acute or chronic disease of any special organ has preceded it. The nervous system and the whole body are often in a condition to which the term 'irritable weakness' has been applied. A severe nerve shock or chronic mental worry belongs in many cases to the causes of general debility. Under the influence of mental depression the breathing and consequently aëration of the blood are diminished; the inclination to take exercise and food is wanting; sleep is disturbed and the nutrition of all the organs and tissues becomes impaired. The treatment is rather similar to that of tardy convalescence, but more difficult, and the failures are numerous. Owing to the long persistence of the debility and the numerous useless trials which have been made, such patients generally have lost confidence in medical treatment, and are not easy manageable. The physician of the spa must exercise judicious authority in insisting on the dietetic and hygienic arrangements which appear necessary to him. If he

uses his influence well, he has a better chance of success than the medical men at home have had.

Anæmia

3. *Anæmia* is of very different import in different persons. The causes vary widely, and the condition of the organs and tissues, and of the nutrition of the body, requires careful consideration in advising the use of spas or other treatment. For the sake of convenience we may divide the anæmias and anæmic invalids into different classes. (*a*) Those who suffer from anæmia caused by direct loss of blood or of the component parts of blood (for instance, when the anæmia is due to hæmorrhage from operations, traumatic injuries, or to purulent, muco-purulent, or serous discharges), are mostly adapted to treatment by iron, and we have to consider whether it is best to use pharmaceutical remedies or chalybeate spas (Chapter XII.), whether pure or compound chalybeate waters are best suited, and whether the spa should be one of high, medium, or slight elevation above sea-level. (*b*) In forms of anæmia caused, not by direct loss of blood, but by acute or chronic disease, neuralgia, different kinds of worry, or sleeplessness and inability to take food, the mildest thermal treatment, combined with forest or mountain climates of moderate elevation, or the latter alone, is often all that can be advised if the person is of delicate constitution; while in others who are less feeble, according to individual conditions, common salt waters, with or without iron, or the gaseous tepid salt baths of Nauheim and Oeynhausen, or the stronger influences of sea air and sea baths, are useful. (*c*) If, as often is the case, anæmia is the result of sluggish portal circulation, constipation, hæmorrhoids, or congestion of the pelvic organs, one of the common salt waters with a certain amount of iron, such as Kissingen and Homburg, or one of the cold sulphated

alkaline waters, like Franzensbad and Elster, must generally precede the use of pure iron or iron and arsenic waters (Chapter XIII.). (*d*) The anæmic conditions produced by long residence in hot climates, often complicated with malarious affections, with enlargement of the spleen and liver, require treatment similar to that of class (*c*); but in these cases it is especially important to select localities which are entirely free from malaria, and where the temperature is moderately cool. Tarasp and St. Moritz in the Engadine offer great advantages. A long stay at high elevations ought always to follow the spa treatment in this class of cases, if possible at places in the immediate neighbourhood of great glaciers, such as Pontresina, the Eggischhorn, the Bel-Alp and Montanvert above Chamonix. (See p. 46.) The spending of several hours every day on the glacier is especially useful.

Chlorosis

Chlorosis we may regard as a variety of anæmia mostly connected with the development of the sexual organs and functions, principally occurring in females. In many cases, in addition to hygienic and dietetic management, rational home treatment is sufficient. There are, however, persons who do not bear the ordinary pharmaceutical iron remedies, and who are much more benefited by pure chalybeate mineral waters. Other chlorotic patients are not at all benefited by iron alone, while they improve rapidly at muriated spas with or without iron, or at the alkaline sulphated spas of Franzensbad, Marienbad, and Tarasp. This is especially the case when sluggish portal circulation is a complication. Arsenical waters are often beneficial when iron waters fail to be so.

4. *Strumous*[1] *and tuberculous affections.*—In former

[1] By 'strumous affections' we here mean those affections of the lymph glands, skin, &c., which were usually termed 'strumous or scrofu-

Strumous and tuberculous affections

years Kreuznach, Ems, Soden, Reichenhall, and other localities were frequently recommended; but we now regard spa treatment as of secondary importance. During summer the sea coasts of England are infinitely more useful to scrofulous children than hot inland spas, such as Kreuznach and Ems; and the education of scrofulous children entirely at the seaside is one of the most successful means of managing such cases. What were formerly called strumous affections of the joints and bones have, since Koch's discovery of the tubercle bacillus, been admitted to be of tuberculous origin, and it has been found that they are in most instances amenable to aseptic surgical treatment.[1] These good results of operation are apparently more readily obtained in suitable climates—for instance, at the hospital of Samaden in the Upper Engadine under the direction of Dr. Bernhard, and at the Royal Sea Bathing Infirmary of Margate.

Pulmonary tuberculosis is now generally admitted to do best under dietetic and hygienic management in what may be termed aseptic climates, in elevated regions, the desert, or on the high seas. Occasionally, however, arsenic spas may be of temporary assistance, such as Mont Dore and La Bourboule. The sulphur waters of the Pyrenees have an old-established reputation, and it is a matter of general experience that the catarrhal conditions associated with pulmonary and laryngeal tuberculoses are often alleviated at Eaux-Bonnes, Cauterets, Le Vernet, Amélié-les-Bains, Bagnères-de-Luchon, and Bagnères-de-Bigorre.

lous,' before they were known to be due to the tubercle bacillus; the course of these affections separates them clinically from tuberculosis occurring in the lungs.

[1] *Tuberculous Disease of Bones and Joints.* By W. Watson Cheyne. Edinburgh and London, 1895.

The hydro-therapeutic treatment, especially by means of cold douches,[1] is only of secondary importance. We may concede to it a strengthening effect on the skin, and the inducement of deep inspirations; but these effects can also be produced by other means, especially by different kinds of respiratory gymnastics.

In some forms of chronic enlargement of the lymphatic glands of the neck, with or without hypertrophy of the tonsils, the use of muriated springs internally and in the form of baths, or of the gaseous thermal 'Soolbäder,' may be resorted to with advantage, if residence at the seaside or pharmaceutical treatment has not been successful. In many of these cases, however, surgical treatment ought not to be deferred too long.

Syphilis

5. *Syphilis* is only to a very limited degree amenable to balneo-therapeutic treatment; ordinary methods of treatment are required. The idea that hot sulphur or other thermal waters can cure it, is without foundation. Hot baths, however, can assist the ordinary medical treatment, especially the mercurial treatment; and energetic courses can be much better arranged abroad, away from home work and home surroundings. Thus gradually a speciality in the management of such courses of treatment has been developed at some places, especially at Aix-la-Chapelle, and the results are mostly satisfactory.

In some persons, however, the constitution becomes entirely undermined by the poison of the disease. Gradually, with or without local affections of the brain and other organs, a cachexia is developed, which is not cured, but may at times be even increased by the usual anti-syphilitic remedies. In such conditions, with widely

[1] Brehmer at Goebersdorf introduced this at the same time that he introduced the open-air treatment of pulmonary tuberculosis.

varying symptoms, the influence of forest and mountain air is often beneficial; but this must be continued during many weeks and months, and may be assisted by the use of simple thermal waters or thermal sulphur waters, especially those at high elevations; for instance, Barèges, Cauterets, Bagnères-de-Luchon, and Wildbad-Gastein. Now and then also iron and arsenic waters find their application in such cases. Judicious hydro-therapeutic treatment is likewise occasionally of great use. The winters ought to be spent in mild climates, by which the diminished powers of the constitution are not overtaxed.

Chronic metallic poisoning

6. Balneo-therapeutic treatment is occasionally resorted to in cases of *chronic metallic poisoning*, especially from *mercury* and *lead*; but the benefit to be derived from it is limited. The most rational treatment is to endeavour to introduce into the blood and tissues substances which form soluble solutions with the poisons deposited in them, and thus assist in their gradual elimination. We scarcely think that mineral waters fulfil this demand. Something may be done by increasing the secretions and excretions, and by thus promoting the removal of the poisons from the body. This object can to some degree be obtained by the internal and external use of indifferent thermal and weak sulphur waters. The further consideration that these poisons are predominantly deposited in the liver leads to the use of the sulphated alkaline and muriated sulphated waters, especially the thermal ones, such as those of Karlsbad and Brides-les-Bains, by which the secretion of bile is stimulated.

Lead paralysis demands, in addition to the ordinary use of the thermal waters, that of douches, massage, and electricity.

7. *Malarial affections* are common amongst those who reside, or have resided, in malarious districts, especially in hot climates. Balneo-therapeutic treatment is only of secondary importance. Pharmaceutical remedies, combined with long residence at high elevations free from malarial air, and especially near glaciers, produce in the majority of persons the most satisfactory results. Malarial cachexia

Occasionally we meet with rebellious cases, complicated, for instance, with catarrh of the bowels, generally with pale motions, in which gentle courses of the thermal muriated-sulphated, or sulphated-alkaline waters, such as those of Brides-les-Bains and Karlsbad, are useful; or, in the case of very delicate persons, the simple thermal waters of Plombières. When there is considerable enlargement of the spleen and liver we must not expect complete reduction of these organs, but we have seen fairly good effects from the sulphated alkaline waters, especially in an elevated locality, like Tarasp. A long stay at high alpine localities ought always to follow the course of waters. (See p. 46.)

When there is no special complication in the abdominal viscera, but only malarial cachexia with anæmia, the waters and climate of St. Moritz or of Ceresole Reale can be recommended. When there are frequent attacks of neuralgia or rheumatism, the simple thermal waters in alpine valleys, such as at Wilbad-Gastein, and the arsenical waters in higher localities, especially at La Bourboule and Mont Dore, deserve a trial, or those of Val Sinestra, which may be employed at Tarasp.

8. *Diabetes mellitus; glycosuria.*—We cannot enter on the pathology of glycosuria, but refer to the works of Frerichs, Seegen, Pavy, W. H. Dickinson, and the last author on the subject, C. von Noorden. In former Diabetes. Glycosuria

years mineral waters were considered pre-eminently curative, and particularly the thermal alkaline, muriated alkaline, and alkaline sulphated waters, amongst which again those of Vichy, Neuenahr, and Karlsbad enjoyed the greatest reputation. Their application, however, is limited. No complete and permanent cure of established diabetes is known to us from spa treatment; but fair results, such as great temporary improvement, which were regarded as cures, have often been obtained, and are frequently obtained now. These results, however, if we examine them without prejudice, are only partially due to the mineral waters. The management of diet, attention to muscular exercise, and regimen in general, have the greater share, and the patient much more readily submits to such management at a spa, away from home. The whole hygienic arrangement greatly assists the diet, and the mental rest and climatic influences are of great importance. For practical purposes we may divide diabetics into three classes, though there is no strict line of definition between them.

(*a*) In the grave and often acute forms in young persons, with excessive loss of sugar, great quantities of urine of high specific gravity, burning thirst, rapid emaciation, and great loss of power, mineral waters exercise scarcely any useful influence, and the fatigue of the journey to a spa is often deleterious, while dietetic and pharmaceutical treatment at home, or in the neighbourhood of home, can often do much to check the progress of the disease. It is to lean cases of this class that the term diabetes is confined by some authorities, such as Dr. Lauder Brunton ('Clinical Lecture on Diabetes,' *St. Barth. Hosp. Journ.* Feb. 1896, p. 67), the well-nourished and fat cases being classed as gouty or fatty glycosuria. If the term 'diabetes' be accepted in this sense, then one must say that only cases of 'glycosuria'[1] are suited to spa treatment. It is because cases of 'chronic glycosuria' or 'benign diabetes' may transform themselves into cases of 'grave diabetes,' that Dr. Pye Smith (Clinical Lecture, *Guy's Hospital Gaz.* March 14, 1896, p. 126) and others prefer to retain the term 'diabetes' for the benign cases, which are in fact the only curable ones.

(*b*) In the chronic forms, without much change of the body weight, either by increase or loss, and without the prominent symptoms of acute diabetes, the proper regulation, without too severe restriction, of diet, and the arrangement of exercise and of the whole manner of living, assisted occasionally by pharmaceutical treatment, will do as much good as mineral waters. We all know that mental shocks and worry have a great share in the production of this form—in fact, of all forms—of diabetes, and that every kind of worry and anxiety must be avoided as much as is possible. Hence frequent changes of climate and locality exercise a beneficial effect in this class of cases, even without any other treatment, spa treatment included, provided the necessary diet can be obtained. If the general strength is, as it often is, diminished, the changes must be of long duration and to milder climates, or to places where the demands on the organism are moderate. The bracing effect of long residence in elevated alpine and forest regions during summer is especially useful, while during winter milder and more sunny localities are to be selected. Owing to the great

[1] By those who use the term 'diabetes' in its more *extended* sense, the term 'glycosuria,' when employed in contradistinction to 'diabetes,' would be applied to such cases only, in which the presence of sugar in the urine is due to temporary causes, and soon passes off without special treatment.

influence exercised by the psychical nervous system in diabetes, agreeable mental occupation is to be combined with the change if possible; hence a visit to Egypt, with a tour on the Nile, or to Sicily or Magna Græcia, or yachting in the Mediterranean, or a stay at Rome or the Riviera, is often attended by great benefit.

In many cases a course of waters during the summer months forms a useful mode of change. In weakly persons the simple thermal baths at spas of moderate elevation, such as Gastein, Buxton, Wildbad, Schlangenbad, and Ragatz, assisted by the internal use of very moderate quantities of muriated alkaline waters, act beneficially; in other cases the internal use of alkaline or muriated alkaline waters alone may be recommended (Vichy, Neuenahr, Obersalzbrunn, Royat, or La Bourboule). It must be understood that a suitable, but not too restricted, diet is essential, and ought to become a habit with diabetic persons. Some cases of this class, as of the next, are intimately connected with a gouty tendency, and this must be taken into consideration in the treatment.

(*c*) In a third class of glycosuria we find the tendency to accumulation of fat, combined with the presence of sugar in the urine, sometimes in small, at other times in large, quantities. Occasionally in this class of cases attacks of uric acid gravel alternate with attacks of glycosuria, and sometimes sugar and uric acid sediment may occur at the same time. Albuminuria sometimes appears in this class of cases, first at intervals, afterwards regularly, but mostly the albumen is present only in small quantities. The glycosuria of fat persons is usually allied with abdominal venosity, and the muscles of the heart are generally weak. It occurs not rarely in gouty persons or persons belonging to gouty families.

In this class mineral waters are frequently useful, especially the sulphated alkaline (Chapter X.), and also the pure alkaline (Chapter VIII.), and the thermal more so than the cold springs. Karlsbad, Brides-les-Bains, Vichy, and Neuenahr owe their reputation in diabetes principally to this class of cases. Contrexéville, too, may be recommended with advantage, especially in those persons who have attacks of gravel alternating with glycosuria. Harrogate and Llandrindod may likewise be occasionally advised, as the small amount of sulphur forms no contra-indication. In this class of glycosuria massage and Swedish gymnastics may be combined with the spa treatment, when the patients are too fat or disinclined to take sufficient ordinary exercise.

Diabetes insipidus may be considered a polyuria of nervous origin, and will be mentioned under the head of disorders of the nervous system.

Urinary gravel

9. *Gravel* is often considered as a disease of the urinary organs, but this is not more correct than to say that diabetes is. It is caused by a disorder in the assimilation of food and in the metabolism of the tissues. There are different kinds of gravel, viz. (1) uric acid and uric acid salts; (2) oxalate of lime; and (3) phosphatic gravel (phosphate of lime and triple phosphates). We may almost restrict ourselves to the first kind, uric acid gravel. Although it occurs from heredity, frequently associated with gout[1] or rheumatism (the

[1] In this connection it may be noted that E. Pfeiffer (*Berl. klin. Woch.* 1896, p. 248) thinks that thermal baths may be useful in deciding whether uncertain pains and joint troubles are gouty or not. According to him, after about twenty thermal baths, such as those of Wiesbaden, the daily amount of uric acid excreted in the urine is sometimes very much diminished (by its half or more), and in such cases he thinks the diagnosis of the uric acid diathesis can be made, and the symptoms be considered gouty. The same phenomenon, he thinks, cannot be observed when some other cause is at the root of the trouble.

arthritic diathesis), or sometimes from drinking calcareous water, or from unknown causes, we mostly find it in persons who eat largely or take much alcoholic stimulants, and do not take enough active exercise. The arrangement of diet and regimen is therefore essential.

In those who are inclined to stoutness and redness of face the sulphated and sulphated alkaline waters (Chapters XI. and X.); in many persons, especially if they are pale and inclined to diarrhœa, the pure alkaline waters (Chapter VIII.) are useful; in lean persons the muriated waters (Chapter VII.) are preferable. Many French physicians prefer the earthy waters of Contrexéville.

As a dietetic beverage the waters of Luhatschowitz (a tumblerful night and morning, either cold or hot) can be recommended with a fair chance of success. In some persons a smaller quantity is sufficient, in others a larger quantity is required. The regular examination of the urine must decide this question. In other persons potions of hot water night and morning, or of the so-called table-waters of Apollinaris, Roisdorf, Selters, etc. (Chapter XVI.), combined with a well-arranged manner of living, are sufficient to prevent recurrence of uric acid gravel.

In persistent *oxaluria* with dyspepsia, the digestion and general metabolism, which are at fault, may often be better rectified by spa treatment (p. 307) in conjunction with arrangement of exercise and diet, than by ordinary drug treatment.

Rheumatism

10. *Rheumatism* requires very different kinds of treatment in different persons and constitutions, and the term, as used at present, includes affections of different nature, which, with increasing knowledge, will become more and more defined from each other. It is scarcely necessary to say that we here occupy ourselves only with the chronic and semi-chronic forms. The nearer a case lies to the acute or subacute forms and stages, the greater

is the care required with regard to the balneo-therapeutic treatment.

Convalescence from acute articular rheumatism

This care is especially necessary with persons convalescent from acute articular rheumatism, or with a tendency to attacks of this disease. The whole system is over-susceptible, the skin is very weak, the circulation is very easily excited, the digestion is apt to be deranged, and a comparatively slight cause may set up a kind of relapse or a fresh attack. Of cases with a tendency to acute articular rheumatism, it may be said that the nearer a person is to a previous attack of the disease and the younger the individual, the more easily a fresh attack may be caused by imprudent management. It is as much as certain that acute articular rheumatism is entirely different in nature from ordinary chronic muscular or articular rheumatism; and the disease is placed here only for the sake of convenience and of the name.

Ordinary balneo-therapeutic procedures are not suitable during the first period of convalescence from acute articular rheumatism; but if the convalescence is very slow, if the heart remains weak and irritable with or without valvular complication, and if travelling, with special care, is permissible, a very cautious course of bathing at the thermal gaseous 'sool baths' of Nauheim is likely to be eminently beneficial.

If the general condition is good, the heart free, and the skin not very weak, but the joints remain more or less stiff and swollen, the simple thermal waters and the muriated waters (Chapters VI. and VII.) are applicable. Douches, massage, and Swedish gymnastics are helpful, but the management of these accessory means requires great cautiousness in the beginning.

The duration of the courses of bathing must often be five or six weeks, and sometimes more, and ought to be followed by a stay at a moderately elevated, sunny, and dry locality, such as Les Avants, Glion, St. Beatenberg, Gurnigel, Badenweiler, or at a fairly warm marine spa.

Chronic articular rheumatism

Chronic rheumatism of joints.—If attacks of rheumatism have left chronic swelling or stiffness of joints without special affection of the whole constitution, the balneo-therapeutic means are numerous. More or less all the hotter simple thermal springs and the thermal sulphur spas can be rendered useful, and also the brine baths ('Soolbäder') in England and abroad. Massage, gymnastics and douches —all with careful adaptation to the individual case—assist the treatment. In many instances local hot applications, like cataplasms of moor earth, peat, simple or sulphur mud, are used with advantage; and also local baths of the same substances or of hot sand.

Chronic muscular rheumatism

In chronic muscular rheumatism the use of hot general baths may be employed as in the preceding group, and can be more freely assisted by douches. In many instances internal courses are required to assist the treatment by baths, according to the nature of the complications, such as dyspepsia, constipation, or the uric acid diathesis. For after treatment in this as well as in the previous group it is necessary to select dry and sunny localities, at moderate elevations if possible, on the southern or western slopes or terraces of mountains. Chronic *lumbago* comes under the same considerations.

Lumbago

The great tendency to relapses in chronic rheumatism requires 'strengthening of the skin' by hydro-therapeutic procedures, by active exercise, and by being much in the open air, which, in so-called rheumatic persons, ought to become parts of their daily life and habits.

Sciatica and similar affections

11. *Sciatica*, *brachialgia*, and similar affections are often more gouty than rheumatic. There are also other causes of these affections. The severer forms are generally due to neuritis, and require at first rest; but

after the acute stage is over simple thermal, or thermal muriated, or thermal sulphur, or cold muriated waters heated, are useful, and the use of the hot bath is often advantageously combined with douches and massage. What has been already said, when speaking of chronic rheumatism, concerning the internal use of waters and the after-treatment, holds good also for sciatica.

Rheumatoid arthritis

12. *Osteo-arthritis* (rheumatoid arthritis, arthritis deformans, or chronic 'rheumatic arthritis') is different from either rheumatism or gout. Cases of this kind are sent to a number of spas, but none exercises a really curative effect. The general condition is, however, frequently improved by very gentle treatment at the simple thermal, thermal sulphur, and thermal muriated waters, occasionally also by the muriated alkaline waters when the spas are situated in good climates, and especially when they are situated at moderate elevations. Change of locality forms an important element in the treatment of this complaint, especially change to dry and sunny climates without excessive heat or cold. Judicious changes of climate and arrangement of diet and of exercises of various kinds, pursued during many years, often give a satisfactory result, though not amounting to a cure.

Senile hip-joint

Senile hip-joint disease (malum coxæ senile) is by the patients themselves usually called gout, sometimes rheumatism. It does not belong to either class; but is generally by medical men classed as a variety of osteo-arthritis, though the pathology, at least of the severer cases, and of the similar affection of the shoulder-joint, is perhaps still open to doubt. In regard to balneo-therapeutic treatment we refer to what has been said on osteo-arthritis.

The multiple nodosity of joints of the fingers is likewise rarely cured; and some invalids have made the round of

Nodosity of finger joints

many spas, on their own or their friends' advice, without being able to point to an improvement in the joints, though their general health had been possibly benefited, more by some one of them than by the others. It is, indeed, the general health which is to be principally considered, and often, through improvement of this, the progress of the local disease is arrested, sometimes for many years, and even the nodosity is occasionally greatly reduced.

Gonorrhœal rheumatism

13. In gonorrhœal rheumatism there is nothing special with regard to balneo-therapeutics. If, after the cure of the exciting cause by ordinary treatment, swelling and stiffness of joints remain, the condition may be treated similarly to chronic articular rheumatism (p. 302).

Gout

14. *Gout* is of very different import in different individuals. Spa treatment is in numerous gouty persons absolutely unnecessary, excepting in so far as many people can be induced at spas to carry out the necessary rules of diet and regimen, while they object to them at home.

On the other hand, it may be said that according to the different states of constitution, and according to existing complications, almost all the spas and climates of Europe may be rendered now and then very useful to gouty people. It would take a little book to exhaust this subject.

In many weak persons only climatic changes and the simple thermal spas (Chapter VI.), the muriated (Chapter VII.), and muriated alkaline waters (Chapter IX.) can be used, and these only in a cautious manner, especial care being taken in old and debilitated persons and in those affected with arterio-sclerosis.

In so-called plethoric constitutions with abdominal venosity and tendency to obesity, the sulphated waters

(Chapter XI.) or the sulphated alkaline waters (Chapter X.) exercise the best effect, provided a long rest in alpine or sub-alpine climates can follow the use of the waters. The alkaline sulphated waters are especially useful in plethoric cases with uric acid deposits. If the tendency to uric acid exists without marked abdominal plethora or congestion, a course of Contrexéville is considered most beneficial by many, especially French physicians; and again by others the alkaline group (Chapter VIII.). In many gouty persons, without pronounced obesity, and with sluggish intestinal action, the muriated waters (Chapter VII.) are preferable to the sulphated. The combination of sulphur with muriated waters as at Harrogate, Llandrindod, Aix-la-Chapelle, Uriage, etc., forms no objection to their being recommended in gouty persons, with or without abdominal congestion. In gouty eczema the muriated alkaline (Chapter IX.) and the sulphur waters (Chapter XIV.) are useful, and amongst the former, Royat has established a reputation, amongst the latter, Schinznach; frequently, it must be added, the cure is not permanent. This, however, is more or less the case with all gouty complaints, especially if the gouty persons relapse into their faulty habits.

Obesity

15. *Fat persons* often think that a course of waters can reduce them to average size; but this is rarely the case without a strict arrangement of diet, exercise (ordinary, or by Swedish gymnastics or massage), and general regimen, which might in most cases be done at home. The sulphated (Chapter XI.) and sulphated alkaline waters (Chapter X.) are, however, able to assist the dietetic and other treatment.

Climacteric changes

16. Although the seven ages of Shakespeare cannot be always distinctly recognised, and are not sharply separated from one another, we all must acknowledge different

stages in the lifetimes of men and women; and the passage from one to another is in some persons attended with more or less serious troubles. The equilibrium of such invalids is easily disturbed, and slight injurious circumstances may produce effects altogether out of proportion to the cause. It is not necessary in the majority of instances to resort to spa treatment, but change of air is mostly beneficial; and occasionally this can be advantageously combined with treatment at simple thermal (Chapter VI.), or at chalybeate (Chapter XII.), or arsenical spas (Chapter XIII.), especially when the spas are in elevated situations.

Premature old age

17. The senile degeneration of tissues, organs, and functions manifests itself in different persons at different ages. In advanced or old age almost all the tissues are inclined to waste, and the functions become inactive. This is especially the case with the involuntary muscular tissues and the organs of the circulatory system, principally composed of them; and as the nutrition of all the tissues and organs depends on the health of the nutritive blood-vessels, the decay of the latter leads to deterioration of the different organs and functions. This shows itself in some persons more in one system, in others more in another. The tendency to early decay of the one or the other system is often hereditary, and can be recognised in it long before old age is due in the natural course of events.

Many persons ask for advice about what they think to be disease, while it is in reality only a manifestation of premature old age in one or another part of the body.

Much can often be done to ward off old age and prolong the life of the whole organism, and of the particular part of the body which shows the tendency to early decay. We all have the opportunity of watching this in the muscular system, in the joints, in the skin, in the urinary organs, in the general nutrition of the body, and in the brain functions.

The means to prevent and, to some degree, to remove premature old age of the whole organism are to be found more in other management than in balneo-therapeutics. The judicious arrangement of diet, general regimen, and especially of exercise and occupation, has a very powerful influence. Not rarely, however, the simple thermal spas (Chapter VI.) can be used with great advantage, especially those situated at higher elevations, and the benefit to be derived from them may be greatly increased by the assistance of Swedish gymnastics and massage and gentle climbing exercise; and if this cannot be had, by respiratory exercises which indirectly act on the heart and entire circulation.

Diseases of the Digestive Apparatus.

18. *Dyspepsia* is a term applied to numerous different conditions, which are associated with disturbance in the digestion of food. Dyspepsia

Dyspepsia is in many persons only a manifestation of a *weak mucous membrane*, in which participate more or less all the mucous membranes of the body, and which again is intimately allied to and forms part of a weakness of the nervous system. The majority of such invalids are thin, and have little resisting power. Mental or bodily exertion is apt to produce or to aggravate the dyspeptic troubles. It often forms a prominent part of neurasthenia. No energetic spa treatment is suitable to such invalids, but the general management may be beneficially assisted by the simple thermal waters (Chapter VI.), especially those in elevated situations; sometimes by a very gentle use of the muriated waters (Chapter VII.) or the muriated alkaline waters (Chapter IX.), followed by a long stay at mountain health resorts.

Dyspepsia alcoholica is the result of a specific catarrh of the mucous membrane of the digestive system, which requires above all things abstinence from its cause, namely, alcoholic beverages, or, at all events, the greatest moderation in their use. The catarrh itself and the complications, if this primary condition is fulfilled, are ameliorated and gradually cured, according to individual indications, by sulphated alkaline (Chapter X.), muriated alkaline (Chapter IX.), or muriated (Chapter VII.) waters, followed by residence at moderate elevations.

The *dyspepsia of smokers* seldom requires balneotherapeutic treatment, but the waters just mentioned are sometimes helpful.

For the numerous forms of *dyspepsia in gouty persons* we refer to what has been said on the management of gout (p. 304).

The dyspepsia attendant on *habitual constipation from torpor of the intestines* is to be removed by the treatment of the latter. The dyspepsia of *anæmia* usually comes under the head of this affection, when there is no gastric ulcer as a cause.

The dyspepsia of the *early stage of pulmonary tuberculosis* requires the climatic and dietetic management of this affection (p. 292).

We might mention some other forms of dyspepsia, but they will either be mentioned under other heads, or will find their treatment in the removal of the noxious causes or habits.

Abdominal venosity

19. '*Abdominal plethora*' (abdominal venosity) is frequently associated with tendency to obesity, and

occurs often in stout gouty persons. In such cases it is mostly treated with benefit by the sulphated (Chapter XI.) and the sulphated alkaline waters (Chapter X.). In only moderately nourished or lean subjects the muriated (Chapter VII.) or muriated alkaline (Chapter IX.) waters are preferable. The sulphur waters with common salt, such as Harrogate, Llandrindod, Aix-la-Chapelle, and Uriage, likewise often exercise beneficial effects. Residence at moderate elevations ought always to follow. Muscular exercises of different kinds are important, and restriction as to food and fluids—the latter especially at meal times—is mostly essential.

Habitual constipation

20. *Habitual constipation* in stout persons requires treatment by sulphated or sulphated alkaline waters, while in leaner persons the muriated waters often produce improvement, though this is not rarely only temporary. The treatment must be continued for a month and often longer, and must not be left off suddenly, else hæmorrhoids may be produced. Fair trial ought always to be given to arrangement of diet and exercise before spa treatment is resorted to. Massage and gymnastics are likewise often preferable to the latter. We shall return to this subject under the head of 'headache.'

Hæmorrhoids

In *Hæmorrhoids*, as far as balneo-therapeutic treatment is concerned, we must likewise consider whether an invalid is stout or lean, and use similar treatment to that recommended in habitual constipation. Alteration of regimen is often of considerable utility.

Catarrh of the intestines

21. *Catarrh of the intestines* may be caused by different conditions, and the treatment must vary accordingly. If it is the result of habitual constipation it requires the treatment of this affection. If it is caused by a weak mucous membrane and shows itself by

Chronic diarrhœa

habitual diarrhœa or frequent attacks of diarrhœa, the management suggested under dyspepsia from weak mucous membrane (p. 307) holds good. The amount of waters taken internally ought to be very limited, and the whole course, including diet, ought to be carefully superintended by the spa physician. The waters and baths of Plombières have acquired a great reputation in this class of cases.

Tropical diarrhœa. Hill diarrhœa

If the diarrhœa is connected with *malaria and dysentery* and consists in frequent pale motions, possibly sometimes containing blood, with imperfect secretion of bile, the spa treatment can take only a small share in the management. This is the case especially with the affection called by Indian doctors *hill diarrhœa.* It frequently happens that the internal use of waters is at first worse than useless, and that only the simple thermal baths at the higher elevations are permitted, while in old cases the muriated alkaline (Chapter IX.), the simple muriated (Chapter VII.), the sulphur (Chapter XIV.), and in rarer instances the sulphated alkaline waters (Chapter X.) can be recommended. The doses of alkaline waters ought to be very small, and the spa treatment ought always to be followed by long residence at high localities with a dry soil. We ought to add that in some very chronic cases with well-marked anæmia, arsenic waters (Chapter XIII) have proved beneficial.

Gastric ulcer

22. *Chronic ulcers of the stomach or duodenum* require dietetic treatment, but the dietetic treatment can be occasionally assisted by the most careful use of alkaline (Chapter VIII.) and warm sulphated alkaline waters, especially those of Karlsbad.

Congestion of the liver

24. Congestion and enlargement of the liver from abuse of alcohol or from malarious affections, or from sluggish portal circulation, or from dilatation and imperfect contractions of the heart, must be treated according

to the most prominent cause; but in almost all cases the judicious use of alkaline (Chapter VIII.) and sulphated alkaline waters (Chapter X.), if the individual is stout, and of muriated waters (Chapter VII.), if the invalid is thin, acts beneficially. The congestion, consequent on dilatation and imperfect contraction of the heart, requires especially careful management, similar to that mentioned under dilatation of the heart (p. 316).

25. *Gallstones* and allied affections, as inspissated bile and bile sand, are generally benefited by the sulphated alkaline (Chapter X.) and the alkaline waters (Chapter VIII.), especially those of higher temperatures. The earthy waters (Chapter XV.), which can be taken in very large quantities, exercise likewise often beneficial effects, principally by the washing out of the small ducts. All these waters dilute the bile and seem to counteract catarrh of the ducts, which has a share in the causation of biliary concretions. In impacted gallstones mineral waters cannot take the place of surgical management. Gallstones and allied affections

26. *Cirrhosis of the liver* in the earlier stages is often greatly benefited by the same treatment which has been suggested under the head of congestion of the liver. In the more advanced forms of the disease spa treatment can only have a palliative effect. Cirrhosis of the liver

We need hardly allude to those cases in which syphilis is the cause of cirrhotic changes in the liver, and in which ordinary spa treatment is more or less useless. (See the section on Syphilis, p. 293.)

27. *Jaundice*, in its very chronic forms, is not rarely treated at spas, but is benefited only in those cases which are connected with catarrh of the bile ducts; these cases require a similar general management to that discussed under the paragraphs (Sections 24 and 25) on gallstones and chronic congestion of the liver, and as regards spa treatment, the same rules hold good. Spa treatment is Jaundice

not applicable to carcinomatous affections, but in some instances it is very difficult to distinguish whether jaundice is due to the more ordinary causes, such as chronic catarrh of the bile ducts and plugging by inspissated bile, bile sand, or biliary calculus, or whether it is due to obstruction in the bile ducts due to cancer. In such cases a very cautious trial of the alkaline or sulphated alkaline waters is permitted, and we know of a number of cases where distinguished physicians had given their opinion for new growth, and where we were ourselves uncertain, when Karlsbad has effected a cure, and thus cleared up the diagnosis.

Ascites

28. *Ascites*, or dropsy of the peritoneum, is one of the affections for which relief is rarely sought from spa treatment, and the proportion of cases in which such treatment can afford relief is only small. When ascites is caused by compression of the portal vein through new growth, or is due to disease of the kidneys, or to tuberculous affection of the peritoneum, balneo-therapeutic treatment ought not to be attempted. When cirrhosis of the liver is the cause of ascites, the cirrhotic affection is mostly far advanced, but in some instances it occurs rather early in the disease, and then the treatment mentioned previously for congestion and commencing cirrhosis of the liver is occasionally beneficial. It is especially in cases where the liver and the portal circulation have been affected through dilatation of the heart, that we have seen some real cures follow spa treatment, mainly from the use of thermal gaseous muriated waters, like those of Nauheim (pp. 83 and 316). Such cures we have seen, not only from the system of combined treatment by baths and exercises, but long ago from the use of the baths at Nauheim, before the exercises had been introduced.

29. *Enlargement of the spleen*, if caused by *lymphadenoma* or Hodgkins' disease, is scarcely ever benefited by spa treatment; and the same may also be said of the splenic tumour of *leucocythæmia* or *leukæmia*. Enlargement of the spleen

The enlargement resulting from enteric fever, erysipelas, septicæmia, puerperal fever, anthrax, and acute tuberculosis, does not fall into the sphere of balneotherapeutics.

The splenic tumour caused by malarious affections is referred to under the head of malaria (Section 7).

The very rare cases of simple idiopathic hypertrophy do not require spa treatment, and for the splenic enlargement connected with cirrhosis of the liver we need only refer to the section concerning the latter affection.

Diseases of the Respiratory Organs.

30. *Chronic nasal* and *naso-pharyngeal catarrh*, if not caused by adenoid or polypoid growth, or syphilitic disease, may be treated by the internal use of muriated, sulphur, and arsenic waters, and by inhalation of the spray of these waters. Cauterets and Mont Dore have acquired considerable reputation, but the cures are rarely perfect. One or more winters spent at dry, warm localities, such as Egypt and the Riviera, and long sea voyages, often do more good than waters, but the latter made use of during summer may be combined with, or rather followed by, the former, during winter. Chronic nasal and naso-pharyngeal catarrh

31. *Chronic bronchial catarrh* or chronic bronchitis, if it is free from complications and merely the result of neglected acute catarrh in persons with a weak mucous membrane, can be treated with benefit by muriated and muriated alkaline waters (Chapters VII. and IX.), especially Ems, Gleichenberg, Baden-Baden, or the Chronic bronchial catarrh

sulphur waters (Chapter XIV.), particularly those of the Pyrenees and Schinznach, or weak arsenic waters, as La Bourboule and Mont Dore. If it is complicated with gout or a gouty diathesis, in weakly individuals the same waters can be employed, while in stout or plethoric persons the alkaline sulphated or muriated sulphated (Karlsbad, Brides) or the stronger muriated waters (Kissingen, Homburg, etc.) may be found useful.

If the chronic catarrh is complicated with a weak and dilated heart, with or without valvular disease, thermal muriated waters, especially in the form of baths (Nauheim and Oeynhausen), are preferable.

In all these cases the balneo-therapeutic treatment ought to be followed by long residence in moderately elevated wooded regions, sheltered from wind, especially near pine forests; for instance, in the Black Forest, the Flimser Waldhäuser in Switzerland, or in sheltered seaside places.

Pulmonary emphysema

32. *Emphysema of the lungs*, as such, is not ameliorated by spa treatment, but the more advanced cases, especially in elderly persons, are almost always complicated with chronic catarrh, and then the treatment by muriated alkaline and thermal sulphur waters, just mentioned, is useful. If the heart is much dilated, attention must be paid to this complication.

Regarding *spasmodic asthma*, we refer to the paragraph (Section 54) under disorders of the nervous system.

Pulmonary tuberculosis

33. *In tuberculosis of the respiratory organs*, as we have already said under the general head of tuberculous affections, balneo-therapeutic treatment has only a slight share. In quiescent cases, however, the catarrhal complications are often much alleviated by the warm sulphur waters at moderate elevations, such as the Pyrenean spas. It is probably to such cases that Eaux-Bonnes

owes the great reputation, which it obtained in former years, in the treatment of consumption. In another class of quiescent and cured cases of phthisis with much emphysema, where there is great delay in the portal circulation, accompanied by dyspepsia or inactivity of the bowels, the very moderate use of muriated (Chapter VII.) or muriated alkaline waters (Chapter IX.), with or without arsenic, has in our experience proved eminently beneficial. In such conditions the waters of La Bourboule, either drunk at the place itself, or at other localities, have been very useful to us. The same may be said of Gleichenberg and its waters, which are little known out of Austria and Hungary, but have a well-deserved reputation in that country.

Diseases of the Circulatory System.

The majority of diseases of the heart and blood-vessels are not to be treated by balneo-therapeutic means. There are, however, some conditions in which these methods can be employed with advantage.

The heart after rheumatic fever

34. The *heart after rheumatic fever* is often in a very weak and irritable condition, especially when the disease has been complicated with endo- or pericarditis, sometimes with both, or with myocarditis, or with all three combined, and when valvular affections have been set up. We have already discussed this subject to some extent under 'Rheumatism' (Section 10), and have there said that ordinary balneo-therapeutic treatment is hazardous in convalescence from any grave form of rheumatic fever. This remark is especially applicable to cases where the heart remains affected after the termination of the acute or subacute disease. Only the most gentle and cautious bath treatment is permissible either at one of the simple thermal spas, or, what is still better, at the

gaseous thermal muriated waters of Nauheim, though the same might be carried out at the similarly constituted, though less known, waters of Oeynhausen. The late Professor F. W. Beneke, of Marburg, who practised during summer at Nauheim, was the first to direct our attention to this subject about twenty-five years ago, and under his guidance many of our heart cases derived very great benefit from courses of baths at Nauheim. In the more recent cases of valvular affection, with decided murmurs from the mitral or aortic valves or both, we have repeatedly witnessed the murmur slowly but entirely disappear, and the heart become enabled to undergo a natural amount of exertion. This was, we may remark, at a time when no exercise cure was practised at Nauheim, and when the place was almost unknown out of Germany.

In more chronic valvular affections the murmur has never disappeared in our experience.

Dilatation of the heart

35. In *dilatation of the heart*, with or without valvular affection, the same class of waters is likewise often useful, and not rarely have we seen the heart become regular in action, the circumference of dulness diminish, the apex beat reappear, and the pulmonary and hepatic troubles, caused by the imperfect contractions, disappear. In a certain number of cases where there was considerable œdema of the legs and effusion into the pleural and peritoneal cavities, these results of the dilated condition of the heart have been entirely removed by long-continued courses at Nauheim. These cures were likewise effected independently of the so-called Nauheim exercises (see pp. 33 and 86) now so much brought before the profession and the public. We ought, however, to add that such good results were not obtained in all cases.

In cases in which the cardiac condition is mainly a

result of, and part of, faulty general nutrition, the treatment has a chance of more lasting success than in cases where the faulty general nutrition is the result of the disturbance in the circulation associated with cardiac valvular defects, adherent pericardium, or atheromatous obstruction in the coronary arteries.

Valvular disease of the heart

36. *Valvular disease of the heart.*—We have spoken of very recent valvular affections under a preceding head (Section 34). As to old-established affections of the valves, spa treatment has no curative influence on them as such; but the morbid conditions to which they give rise, viz., dilatation of the heart, catarrh of the lungs, congestion of the liver and abdominal organs, may be relieved. Concerning balneo-therapeutic treatment of these conditions, we refer to what has been already said under their respective heads.

Fatty and fibroid degeneration of the heart

37. *Fatty or fibroid degeneration of the heart* is a somewhat uncertain matter as far as spa treatment, as well as other treatment, is concerned. Whoever knows Sir Richard Quain's or Dr. Kennedy's writings on the subject, will feel that the first point to be settled—the diagnosis—is very difficult in the earlier stages, and in the later stages of the disease spa treatment is too hazardous. We think, however, that in some cases, where we were tolerably sure of our diagnosis by the irregularity, weakness, and slowness of the pulse, coupled with breathlessness on slight ascents, and with tendency to drowsiness and to syncope, benefit has been derived from the cautious internal use of muriated waters, especially those of Kissingen, several times also from the sulphated alkaline waters of Karlsbad, and twice from the employment of the gaseous thermal salt baths of Nauheim. In some other cases, however, and amongst them some where the diagnosis was later on confirmed by the

examination after death, we were disappointed by the results of balneo-therapeutic treatment.

In some cases, probably of this class, with prominent anæmic complications, iron waters, as well internally as in baths, have acted favourably. We need scarcely say that in this grave affection, spa treatment must be assisted by climate and attention to the diet, exercise, and mental condition of the patient.

Fatty infiltration of the heart

38. *Fatty infiltration of the heart* (to which is sometimes applied the term 'fatty heart') often interferes with the function of the organ. The treatment recommended for obesity is applicable in this condition; but in addition tonics, such as quinine and strychnine, ought sometimes to be employed during the use of the waters, and chalybeate waters are mostly advisable for after-treatment.

Palpitation of the heart

39. *Palpitation of the heart,* without appreciable, permanent structural lesion, is to be regarded as a neurosis, and balneo-therapeutic treatment is only in exceptional cases required. Ordinary home treatment, with attention to the predisposing and exciting causes, can generally do more than spa treatment; in some cases changes of locality and surroundings, suitable climate, and the removal of the worries and the excitement of social life, are the best curative agents. In some persons, however, dyspepsia is the exciting cause, and for this we refer to what we have said on dyspepsia (Section 18). In others, the circulation in the abdominal, and especially pelvic organs, may be attended to at spas. When anæmia and chlorosis form the predisposing conditions, the treatment ought to be directed to these affections. In hysteria and allied affections with over-excitability in one sphere of the nervous system, and failure of the inhibitive power in another, the simple thermal baths (Chapter VI.) in medium elevations often exercise a very beneficial effect;

Schlangenbad has an old-established reputation in cases of this nature.

If palpitation of the heart is due to Graves' disease, perhaps in an incomplete form or early stage, where the whole complex of symptoms is not yet fully developed, the remarks made under the head of this affection can be referred to.

Other functional disturbances of the heart, such as irregularity in the rhythm, great frequency of contractions, and hæmic murmurs, do not, *per se*, come into the consideration of balneo-therapeutics.

Varicose veins

40. *Varicose veins* of different parts of the body, especially the legs, are often connected with venous congestion of the abdominal organs and habitual constipation, and may be treated by the means recommended under that head (Section 20); but mechanical appliances are seldom rendered unnecessary by the use of baths and waters.

Fibroid degeneration of the heart and small blood-vessels

41. *Fibroid degeneration* in the small blood-vessels (arterio-sclerosis or arterio-capillary fibrosis), in the early stages, if it occurs in stout and plethoric persons, is, to some degree, benefited by the alkaline sulphated (Chapter X.) and alkaline waters (Chapter VIII.), and in lean persons by the muriated waters. Perfect cures are scarcely to be expected, and the manner of living in such cases has to be permanently altered, so as to prevent, or, at all events, to retard the further progress of degeneration. The same applies to cases in which early fibroid changes in the heart-walls themselves are suspected; when more advanced changes have taken place, what has been said in the paragraph on fatty degeneration of the heart (Section 37) can be accepted here also. Localised fibroid changes in the heart-wall, such as lead to cardiac aneurism, if they could be detected during life, would, of course, be unsuitable to balneo-therapeutic treatment.

Atheromatous changes

42. *Atheromatous changes*, in their commencement, are to be similarly treated; but in more advanced cases, with atheromatous valvular affections, and when commencing aneurism is feared, the greatest care is necessary in the use of balneo-therapeutic treatment. When there is actual *aneurism*, such treatment is inadmissible.

Angina pectoris

43. *Angina pectoris* is a more debateable matter. Many cases have come before us with the diagnosis of 'angina pectoris.' In some of them the anginoid attacks were clearly due to gastric derangements in persons with a so-called weak heart, mostly a dilated heart. In such cases the management of the diet and regimen, either at home or combined with treatment at spas, such as recommended for dyspepsia or dilatation of the heart (see under these heads), has been mostly successful. In cases when the symptoms were such as to leave a doubt whether there was real angina pectoris or merely functional trouble, the treatment was in many cases likewise successful; but in others only a more or less temporary success—occasionally for years—was obtained, while later on indubitable and fatal angina pectoris manifested itself. In cases where the diagnosis of angina pectoris is certain, it is prudent to avoid spa treatment and long journeys to spas, excepting perhaps in certain cases where the intervals between the attacks are long, and where the presence of gouty or dyspeptic troubles encourages spa treatment. In the latter cases a trial of spa treatment may be made with careful arrangement of the journey and cautious use of balneo-therapeutic means.

Disorders of the Nervous System.

Physicians have, in former years, employed spa treatment in many diseases of the nervous system, but its use in the more serious organic diseases is very limited.

44. *Locomotor ataxy* or *tabes dorsalis* is a disease for which very different spas are frequently recommended, but the benefit derived is mostly only very moderate. The so-called 'lightning pains,' the dull aching pains and the 'pins and needles' are not rarely mitigated by simple thermal baths (Chapter VI.) and by the gaseous muriated baths, amongst which Oeynhausen has obtained a certain reputation in this class of cases. As arsenic appears occasionally of some use in tabes, so some persons affected with this disease have derived a certain amount of good from arsenic waters. Sulphur waters, too, have at different times been recommended, but we cannot say that we have seen more good from them than from simple thermal waters. As syphilis is in many instances a prominent ætiological factor in locomotor ataxy, inunction of mercury combined with the baths at Aix-la-Chapelle has been strongly recommended. We can say that we have repeatedly seen great benefit from this plan, while it has occasionally entirely failed, and we are not sure of a single permanent cure, although the good result has not rarely lasted for a period of several years.

Tabes dorsalis

In some rare instances, which have remained stationary for more than twenty years, the use of thermal baths in summer, and yachting during winter in sunny climates, including the Nile, seem to have acted very beneficially.

45. In the various *forms of paralysis*,[1] excepting those of clearly syphilitic origin, balneo-therapeutic treatment is more or less useless, and occasionally injurious, and in syphilitic cases the treatment at spas is mainly the

Paralyses, &c.

[1] See also p. 50. It is scarcely necessary to say that the false paralysis, due to chronic rheumatism, old injuries to joints, etc., with partial ankylosis, or adhesions about the joints, are not referred to in the above heading.

ordinary specific one. In some cases of old *hemiplegia* the simple thermal baths are beneficial by improving the general health.

In the so-called *apoplectic habitus* in plethoric persons the sulphated waters (Chapter XI.) used internally, and the cold sulphated alkaline (Chapter X.) may be of prophylactic value.

Epilepsy, unless it is caused by cerebral syphilis, ought not to be treated by spas.

In *progressive muscular atrophy* no good results are known to us from spa treatment.

The same may be said of *pseudo-hypertrophic muscular paralysis*, and of other kinds of *primary muscular dystrophies*.

Headache

46. *Headache* has many varieties, and is produced by many different causes, a few of which are to some degree within the scope of balneo-therapeutic means, while others are not. We will discuss some of the principal varieties.

In headaches caused by *organic disease* in the brain or in the cranium, the treatment in question affords no or very little relief, unless the complaint is of syphilitic origin; the limited share of spa treatment in cases of syphilis is mentioned in Section 5.

Anæmic headache

For the headache connected with *anæmia* we refer to the paragraph on the latter (Section 3), and also for headaches caused by long-continued leucorrhœa.

Rheumatic headache

There is a headache of *rheumatic nature*, to which the treatment mentioned under 'chronic rheumatism' is applicable; we may add that in this form massage of the neck and scalp is helpful.

Headache from venous congestion

In a great number of cases the headache is probably due to *venous congestion* within the cranium mostly in connection with a weak venous system, either local or more or less general. The exciting causes are mostly in distant parts of the system, and more frequently within the abdominal cavity, and for the balneo-therapeutic treatment we refer to the sections on these causes.

Headache from constipation

Habitual constipation is a frequent cause of dull headache. It is very probable that the absorption of certain ptomaines, or the imperfect removal of them by the excretory organs, has much to do with these headaches, although we do not overlook the fact that the impediment in the abdominal circulation caused by constipation may have a share in their production. We often find that courses of sulphated (Chapter XI.) and alkaline sulphated waters (Chapter X.) in stout persons, and of muriated waters (Chapter VII.) in lean persons exercise a beneficial effect in the treatment of this form of headache. Karlsbad, Marienbad, Franzensbad, Tarasp, Kissingen, Homburg, and similar spas have obtained a well-deserved reputation. The treatment, however, must not be confined to the course of waters, but the diet[1] and manner of living must be regulated

[1] It cannot be doubted that the absorption of toxins, produced in the alimentary canal (one form of auto-intoxication), often takes a leading share in the production of headaches and irritability, both when constipation is present and when it is not, but especially in the former case. The exact nature of the toxins and fermentations giving rise to them, and the conditions which favour these fermentations, are at present not sufficiently known, though with regard to the question of diet further knowledge on these points is doubtless much needed. Probably excessive fermentation in the bowels is often due, not so much to the exact quality of the ingesta, as to want of vigour in the intestinal walls to hinder it. The peristalsis may be too slow, and the intestinal secretion may be deficient in quantity or quality. The amount of food ingested may be excessive, so that though the intestinal secretions are perhaps normal, they are yet relatively insufficient for the quantity of food. Exercise may remedy some of these states of affairs without any alteration in diet, either by indirectly checking the undue fermentation,

while at home. In many of these cases exercise ought to form an important habit of daily life, and if the subject were properly considered, we should not find the great unwillingness to obey this rule. Exercise (for instance, walking, riding, cycling, and active games) not only acts on the muscles and blood-vessels of the lower limbs and abdomen, but it increases respiratory movements and the amount of oxygen inhaled and absorbed by the blood, and through this favours the oxidation of albuminoid substances and ptomaines; it also strengthens the capillaries and veins not only of the lower part of the body, but also those within the cranium.

Headache from uræmia and cholæmia

There are other forms of headache from imperfect metabolism and excretion, which may be classed under toxæmic headaches, especially those from disease of the kidneys, included in the different degrees of *uræmia*, and of the liver in minor varieties of *cholæmia*. In the pronounced forms there is no possibility of spa treatment, but in the slighter forms mineral waters can influence the excretions, and the sulphated alkaline (Chapter X.) or muriated (Chapter VII.) waters, given according to individual peculiarities, will be found beneficial, whilst in other cases waters acting on the skin (as the simple thermal) may do good.

Headache from alcoholism

A form of toxicæmic headache is that sometimes attending *alcoholism*. The ordinary transitory alcoholic headache cannot come into consideration, but for the

or by aiding the oxidation and excretion of the toxins after their absorption. In other cases a diet may be recommended made less rich in nitrogenous or carbohydrate constituents, or otherwise modified to suit the individual digestive peculiarities of the patient. Often the mere decrease in the total amount of the food, and the eating it more slowly and at more regular intervals, may suffice to check the abnormal fermentation (which may be accompanied or not by dyspeptic feelings) and the absorption of the toxins which lead to the headaches.

chronic form, when connected with dyspepsia or alcoholic cirrhosis, we refer to the paragraph on alcoholic dyspepsia (Section 18). When the chronic alcoholic headache is due to organic changes in the brain or meninges, in addition to removal of the primary cause, climatic treatment and simple thermal waters may effect a certain amount of improvement.

Headache from dysmenorrhœa

The periodic headaches accompanying *dysmenorrhœa* require attention to this complaint. Imperfect circulation in the abdominal organs, frequently combined with constipation, is not rarely the cause of these headaches; hence Franzensbad and Kissingen are so often useful. Sometimes anæmia is a well-marked complication, and renders chalybeate waters advisable, either alone, or to follow the first-mentioned treatment.

Headache from asthma

The headache, which is often associated with chronic *asthma*, is probably due to imperfect aëration of the blood; it disappears mostly with the amelioration of the *asthma* (see Section 54).

Headache from chronic bronchitis and dilatation of the heart

Chronic bronchitis and dilatation of the heart are likewise in some persons attended with headache, which seems to be partly due to imperfect reflux of venous and imperfect supply of arterial blood; but defective oxidation and the retention of ptomaines have probably a share in it. We refer to the treatment mentioned under the heads of chronic bronchitis and dilated heart.

Neurotic headache

There is a kind of headache to which, for want of some better term, we will leave the label *neurotic headache*; it is sometimes one of the sequelæ of exhausting disease, sometimes caused by excessive mental work, chronic sleeplessness, shock, or worry. Some individuals or families are more predisposed to it than others. Change is an important element in the treatment, at first to medium, later on to higher elevations; the first part of this treatment may advantageously be combined with the use of simple thermal waters.

Megrim or sick headache

Sick headaches (Megrim, Migraine, or bilious headaches), for the pathology of which we may refer to the works of Dr. Edward Liveing and Professor P. W. Latham, vary considerably in different persons, and even in the same person at different times. Balneotherapeutic treatment is on the whole disappointing. If there is likewise habitual constipation, benefit is sometimes derived in persons of full habit from the sulphated alkaline and sulphated waters (Chapters X. and XI.), and in lean persons from the muriated waters (Chapter VII.). We can never give a promise in such cases; but occasionally they find great improvement for many months and even years.

Tic douloureux and facial neuralgia

47. *Facial Neuralgia and Tic-douloureux.*—Ordinary treatment, combined with change of air, is generally more applicable than balneo-therapeutic treatment. There are, however, cases in which anæmia is the predisposing cause, and in which the spa treatment, beneficial in anæmia, deserves a trial. In other cases a malarious taint is present, especially in neuralgia of the supraorbital division of the trigeminal ('brow-ague'), in which cases chalybeate spas, and arsenical spas, and long residence in dry localities of high altitude are generally useful. In other cases rheumatism appears to be the cause, and the simple thermal waters (Chapter VI.), occasionally also the iron waters, may be tried. When gout is associated with facial neuralgia, it may require the principal attention. The severest and most typical forms of 'tic douloureux' are not likely to yield to balneo-therapeutic treatment, and indeed they seem to yield permanently to no, not even surgical, treatment.

Clavus hystericus is not a true neuralgia, and we

must refer to what is said in the paragraph on hysteria, for spa treatment in clavus and 'hysterical neuralgias.

48. There are many other neuralgias for which invalids often desire spa treatment. The most frequent are perhaps: *Intercostal neuralgia, occipito-cervical, cervico-brachial, crural*, and *lumbo-abdominal* neuralgia. Some of these so-called neuralgias are in reality due to neuritis. The balneo-therapeutic indications are rather similar to those just mentioned (Section 47). Amongst the neuralgias of internal organs *gastralgia* and *cardialgia*, though they are occasionally, apparently at least, primary, are frequently due to dyspepsia, and we may refer to the paragraphs on that disorder (Section 18).

Other neuralgias

49. *Disorders of sleep* are numerous and of different nature. They often tax the attention of the physician to a great degree, and require his intimate knowledge of the habits of the sufferer, and of the peculiarities of his nervous, circulatory, and digestive systems. They belong more to the sphere of home treatment; but often when this fails, climatic and spa treatment may render good service. We cannot enter into the many causes of disordered sleep, the removal of which often leads to a return to the natural condition; but there are cases connected with the state of the circulation, and others more directly with that of the nervous system, where the ordinary hygienic arrangements in the life at home fail, and where it is undesirable to employ the usual remedies to produce sleep.

Disorders of sleep

We may roughly divide the disorders into two classes, viz., *defective* and *excessive* sleep.

(*a*) The very large class of cases of *defective sleep*, insomnia, includes many varieties, which we cannot fully discuss here. Some persons are unable to fall asleep, others wake a hundred times and oftener during one night, others have distressing dreams, others readily fall

Insomnia

asleep, but awake after some hours, and cannot go to sleep afterwards, and are then worried by thoughts of anxious nature. Many of these cases come under the head of irritable weakness, and are greatly benefited by travelling, by change of air alone, by the use of simple thermal waters (Chapter VI.), notably at Schlangenbad and Plombières, by long residence at moderate elevations, especially in forest districts, with comparatively little movement of air. In some persons dilatation of the heart and defective circulation in the head are the cause, and in such cases, in addition to the means just mentioned, the use of the gaseous thermal muriated waters acts as a promoter of sleep. In cases where *anæmia* is the cause of sleeplessness, the treatment must be directed to that affection. Many scientific workers suffer from insufficient sleep after having continued to work assiduously for some months. The same is the case with many lecturers and professional men. Often this can be prevented by regular daily open-air exercise or by regular breaks in the work. Some sufferers of this class always regain sound sleep by courses of baths at indifferent thermal spas, others simply by residence at high elevations. For the sleeplessness often accompanying asthmatic conditions, even independently of the attacks, we refer to Section 54.

Excessive sleep

(*b*) *Excessive sleep* is often met with in so-called plethoric subjects, especially those with 'abdominal plethora' (Section 19), and in cases with a tendency to cerebral apoplexy. Such persons are apt to fall asleep, as soon as they are left alone, especially after meals, and sometimes, if they are allowed, sleep on for three or four hours at the time, in addition to nine or ten hours of heavy sleep at night. They fall asleep not only while reading, but also while writing a letter, or at the dinner table. These conditions of excessive sleep are usually less distressing to the invalid than those of defective amount of sleep; but they are infinitely more grave. They require, in addition to strict arrangement of food and exercise, the sulphated alkaline waters (Chapter X.), or, if occurring in lean persons, the muriated. After the course of waters the invalids should be sent for a month or longer to moderately elevated localities, where they can spend the greater part of the day in the open air, and should not be allowed to sleep more than seven hours in the twenty-four. The amount of food and beverage must be restricted, and a large amount of exercise or Swedish gymnastics should be an essential part of their regimen when at home.

The amount of sleep which different people require varies considerably, not only at different ages, but at the same age. We do not speak here of childhood and the time before adolescence, but only of the periods of manhood and advanced age. Many persons require only five hours, and ought not to worry themselves if they obtain no more. Most people ought to be satisfied with six or seven hours; in only very few are eight necessary or allowable. But many are not satisfied with this limit, and induce by too much sleep a comparatively early degeneration of blood-vessels, especially the veins and capillaries.

50. *Hysteria*, though the term is faulty, is a real disorder of the nervous system. We do not apply the word to nerve disorders confined to women, and connected with the womb, but, to use the words of Dr. Buzzard in his instructive article in Quain's 'Dictionary of Medicine,' 'to a condition of the nervous system fairly defined, but the intimate pathology of which is not known, characterised by the occurrence of convulsive seizures and by departures from normal functions of various organs, leading to very numerous and often perplexing symptoms. Hysteria

These are apt to simulate those commonly arising from definite alterations in structure, but differ from the latter in the fact, that they may often, even when at their worst, be removed instantaneously, usually under the influence of strong emotion.' Want of will and want of inhibitory power are further prominent symptoms. Spa treatment has no special curative influences, but occasionally can ameliorate concomitant conditions such as dyspepsia, constipation, anæmia (see under these headings). The irritable weakness of the nervous system is often favourably influenced at the simple thermal spas (Chapter VI.), amongst which Schlangenbad and Plombières have obtained a special reputation. The change in surroundings, the intercourse with strangers, inducing mental restraint, the authoritative influence of a new doctor are helpful agents. The regular employment of the day by the drinking of waters and bathing, by the *table d'hôte* meals, by the promenade, and by listening to the music, is a most important element in the spa treatment of many cases of this class. In our experience the improvement was often only transitory, but extended in some instances over many months and years, and was permanent in two long-standing cases, but this was to be ascribed to judicious surroundings, and to the happy influence of successful occupation.

Hypochondriasis

51. *Hypochondriasis*, though the term is justified by old usage, is likewise a defective name, for the disease is of course not in the hypochondria, but in the nervous system. In many cases where no abnormality in the physical state and in the condition of the organs can be discovered, very little can be done by spa treatment, excepting in so far as it supplies a source of occupation. Travelling under favourable circumstances (*i.e.* with judicious friends), and more or less absorbing occupation,

are, indeed, by far the best means of treatment. The cases of hypochondriasis combined with physical disorders, hypochondriasis with a pathological 'substratum,' offer more scope for treatment, and have on the whole a more favourable prognosis. Constipation, hæmorrhoids, dyspepsia, and gouty complications, are to be treated in the way discussed under these headings. Severe courses, however, must be avoided, for the general health of hypochondriacs is easily lowered, but not so easily raised up again. Syphilis may sometimes be present, but it has nothing to do with the disease itself, although some hypochondriacs are incessantly haunted by the idea that they are syphilitic, and demand antisyphilitic treatment, which is mostly much worse than useless.

Neurasthenia

52. *Neurasthenia*, sometimes called 'Beard's disease,' is a term which is often misused, but it expresses much of the condition present in the class of cases so well described and treated by Dr. Weir-Mitchell and Dr. William Playfair.

Spa treatment in itself is entirely useless, but it is sometimes possible to adopt management and treatment during a stay at a spa, which in ordinary home-life is not easily carried out. Weir-Mitchell has shown us that removal from home surroundings is almost indispensable. The two other elements of his treatment, forced feeding and massage, are likewise very important, but not always to the same degree. We have seen occasionally satisfactory results, for instance at Schlangenbad, where the removal from home influence, the authoritative advice with regard to food and exercise and baths, did what isolation and rest in a confined house in town, with massage and forced feeding, did not do. Much depends on the degree of the disease ; in slight cases spa treatment is sometimes very useful, in severe cases it is useless.

There are many relapses, but some invalids are entirely cured. One of the first cases on which Dr. Weir-Mitchell tried his treatment was a lady who had been attended by us, but who remained thin and weak and unfit for the daily demands of life. She was entirely cured by Dr. Weir-Mitchell, and now judiciously superintends the employment of her large fortune for philanthropic purposes.

Exophthalmic goitre

53. *Exophthalmic goitre* (Graves' disease or Basedow's disease), though the thyroid-intoxication and even the infection theories may be partially true, and though the primary ætiological factors may not always be the same indifferent cases, must be considered a disease in which the nervous system is always affected, whether the nervous affection is the primary one or not. The soothing simple thermal (Chapter VI.), and the gaseous muriated baths often give beneficial results in the milder and chronic cases; and when the disease is complicated with anæmia, the chalybeate waters may be useful. The removal from the excitement and worry of home life is helpful in this treatment. Climatic influences are important, and residence in elevated regions is mostly preferable. We may say that we have in several cases seen arrest, even amounting to cure, from migration to, or prolonged stay in elevated regions. The acute and severest cases are, of course, unsuitable to spa treatment.

Spasmodic asthma

54. *Asthma* pure and simple, or true asthma, is a nervous affection; it may be regarded as a neurosis manifested in branches of the pneumogastric nerve.

In the management of pure 'spasmodic' asthma mineral waters are of secondary importance, while climate takes a much greater share. Arsenical waters and sulphur waters are often used, and especially those of Mont Dore, La Bourboule, and the Pyrenean spas; but it is not certain whether the benefit frequently derived is not due rather to the elevated situation of the spas than to their waters; for, although one can never say, before a trial has been made, which climate will suit an asthmatic person, experience is in favour of elevated regions, especially those with little wind; and the younger the individual is, the more likely is he to obtain benefit from long residence in elevated sunny regions, especially in winter, but also in summer. If, however, asthma is combined with advanced emphysema, very high elevations are not to be recommended, but moderate elevations in sunny positions with some shelter from winds, such as Grasse, near Cannes, and places even as elevated as Glion and Les Avants, above Montreux, offer good chances.

In *asthma associated with chronic bronchial catarrh*, the recommendations given under the latter heading hold good, and here again Mont Dore has established for itself a great reputation. In persons, however, with marked dilatation of the heart, in advanced emphysema, and in old people, we have repeatedly seen unsatisfactory results from treatment there, while muriated alkaline (Ems, and Royat, and Gleichenberg) and sulphur waters in lower situations have been more useful, and also the gaseous thermal muriated waters (Nauheim). We have seen marked benefit in several cases from a course of treatment at Weissenburg in Switzerland, and were inclined to ascribe the good result, not so much to the waters, as to the peculiarly sheltered position of the place in a hollow surrounded by high rocks and fir trees, and its moderate elevation above sea-level.

In *asthma complicated with gout*, the latter affection must be taken into consideration; and in asthma connected with affections of the skin or of the abdominal viscera, treatment of these may occasionally cure the asthma.

Diabetes insipidus

55. *Diabetes insipidus* is mostly dependent on the nervous system, and is sometimes an early symptom of degenerative processes in the nerve centres; in other cases it is only functional. Chalybeate waters, simple thermal waters at high elevation, and long residence in alpine regions, are mostly useful. We do not apply the term either to the transitory attacks of frequent micturition from nerve influences, or to that form of polyuria connected with disease of the kidneys.

Cutaneous Affections.

56. Skin diseases were formerly very generally treated at spas, but the benefit to be obtained from balneo-therapeutic treatment is very moderate, though the use of simple thermal baths is useful. We may refer to a short judicious survey of the subject by Dr. Robert Liveing in Quain's *Dictionary of Medicine* (2nd edition). We will glance at some of the principal affections.

Acne

Acne is sometimes complicated with anæmia, and in such cases the balneo-therapeutic treatment of the latter is applicable, though it rarely cures the acne, and the ordinary home treatment, too, is often powerless. The affection almost always disappears later on spontaneously.

Acne rosacea or gutta rosea offers no great scope to spa treatment, excepting that complications, such as dyspepsia, may be treated by mineral waters, and that the removal of these complications often checks the skin affection.

Eczema

Eczema is of all skin diseases the one for which the sufferers most frequently demand balneo-therapeutic treatment, but local pharmaceutical remedies give mostly greater relief. If there is any distinct gouty complication, or constipation, or glycosuria, we may refer to what has been said under those headings. Schinznach, Uriage, and Saint-Sauveur have acquired a certain reputation, and in individuals belonging to gouty families, Royat and La Bourboule. The simple thermal waters, likewise, prove sometimes useful. In very torpid, chronic cases, which in their appearance resemble psoriasis, Loèche-les-Bains exercises very good effects, but these are likewise often only temporary. No certain promise can be given, and the spa physician must proceed cautiously.

Urticaria

The tendency to repeated attacks of urticaria is occasionally very obstinate, and is in many instances as little influenced by spa treatment as by ordinary treatment. There are instances in which a gouty disposition is connected with urticaria, and in which alkaline (Chapter VIII.) and alkaline muriated waters (Chapter IX.) do good; in other cases sulphur waters, such as those of Schinznach and Uriage, are useful. In a considerable number of cases urticaria has a distinctly *neurotic* element in its nature; thus we have seen persons in whom it alternates with attacks of asthma, or with attacks of palpitation of the heart, and again others where mental emotion is apt to bring on an attack. In this neurotic class change of locality and surroundings and escape from worry is important; at the same time simple thermal waters (Chapter VI.) and the arsenic waters of La Bourboule are useful. According to Professor A. E. Wright, of Netley, some cases of urticaria are associated with deficient blood coagulability, and might be treated with calcium chloride, which increases the coagulability of the blood (*Lancet*, January 18, 1896). In such cases a diluted 'Mutterlauge,' rich in calcium chloride, like that of Kreuznach, might possibly be taken with advantage, if Professor Wright's views are correct; the salt-water baths, if not too irritating to the skin, would exercise a beneficial influence on any associated scrofulous or rickety tendency.

Purpura

Purpura.—What has just been said regarding chloride of calcium would apply equally to certain chronic purpuric eruptions, and to children with a hæmophilic tendency.

Lichen

Lichen.—Much of what was formerly called lichen is now classed with eczema. In chronic cases of lichen ruber or lichen planus there is often a certain degree of anæmia, and chalybeate or arsenical waters may be useful.

Psoriasis

Psoriasis is as intractable to balneo-therapeutic as to ordinary means of treatment. The waters mentioned under the heading 'Eczema' are often attended with more or less transitory relief. The long-continued immersions in the warm waters of Loèche-les-Bains (Leukerbad) have, up to this time, led to better results than the treatment at other spas. The bathing in these waters during four to six hours every day produces after some time a superficial cutaneous inflammation, which is mostly followed by complete disappearance of the scaly eruption, but the recurrence of the disease after some months is rather the rule than the exception.

Prurigo

Prurigo, i.e. real prurigo not pruritus, may be regarded as more or less incurable. Loèche-les-Bains claims some successes, but, as far as our limited experience in this comparatively rare disease goes, they are not permanent.

Pruritus

Pruritus is mostly caused by disorder of internal organs, or a gouty condition. Derangement of the liver, especially in its secretion of bile (partly an excretion), is a frequent cause. Karlsbad and Vichy are useful in this variety. In women *pruritus genitalium* is mostly connected with uterine affections or with glycosuria, and ought to be treated accordingly. For the troublesome *pruritus ani*, spa treatment alone is useless, unless it be associated with constipation or hæmorrhoids, in which cases the treatment applicable to those conditions is occasionally curative.

Pruritus senilis (sometimes called 'prurigo senilis'), especially when it occurs in gouty and nervous persons, is often benefited by Schlangenbad, Plombières, and other simple thermal spas (Chapter VI.), also by sulphur waters such as those of Schinznach and the Pyrenean spas, and by passing the winters in warm and dry climates with open-air life.

Seborrhœa

Seborrhœa sicca capitis, or *pityriasis capitis*, in young and old is much more suitable for ordinary than for balneo-therapeutic treatment.

Pityriasis rosea

The same is the case with *pityriasis rosea*.

Pityriasis rubra

Pityriasis rubra in its typical severer forms is doubtless unsuitable for balneo-therapeutic treatment. The simple thermal waters (Chapter VI.) answer to some degree in milder cases connected with much irritation. In very chronic cases without much irritation the sulphur waters (Chapter XIV.) are sometimes useful, especially those of Uriage. Ordinary home treatment ought, however, to be tried first.

Epiphytic skin affections

The different kinds of *epiphytic* skin affections, such as *tinea trichophytina* (ringworm, or tinea tonsurans) and *tinea versicolor* (pityriasis versicolor), are better treated by ordinary means.

Boils

Furuncles or *boils* mostly yield to local and general home treatment, but in very obstinate cases a climatic change is useful, and both seaside and mountain air can be recommended. In some anæmic cases the change is advantageously combined with the use of chalybeate or arsenical waters (Chapters XII. and XIII.), especially at St. Moritz and Ceresole Reale; or of the muriated alkaline arsenical waters of La Bourboule. Sulphur waters, too, have been found useful, especially the muriated sulphur waters of Harrogate and Llandrindod in England, and Uriage in France.

Syphilitic skin diseases

In *syphilitic skin diseases* the treatment of syphilis is the main point, not the skin. Balneo-therapeutic treatment can slightly aid the ordinary treatment, but we need scarcely enlarge on what we have already said under the head of 'Syphilis' (Section 5).

Skin weakness

57. *Weakness of the skin* is a condition to which in general not much attention is paid, but which is, nevertheless, very important. The skin is poorly nourished, is apt to perspire profusely, and atmospheric changes acting on the skin often cause rheumatic conditions or catarrhal affections of the mucous membranes, bronchitis, diarrhœa, or abdominal or facial neuralgia, according to the individual tendency. This state of the skin belongs in many instances to the sequelæ of acute diseases, but in others it is part of a generally weak constitution. In very delicate persons the gaseous thermal muriated baths of Oeynhausen and Nauheim are preferable to stronger measures; in other cases hydrotherapeutic treatment, adapted to the power of the constitution, may be used, and in stronger persons sea baths may be employed. Prolonged change of climate and open-air life are most useful, and well-arranged sea voyages exercise, in the case of 'good sailors,' a most beneficial influence.

Diseases of the Urinary Organs.

These come only to a slight degree under spa treatment.

58. The various forms of nephritis require treatment by ordinary means much more than by balneo-therapeutics; the vapour bath and the hot-air bath, and even hydro-therapeutic means (the hot bath and the pack), are in these affections used in hospital and home treatment, combined with other means.

Disease of the kidney. Bright's disease

In the early stages of chronic interstitial nephritis, occurring in gouty subjects, benefit may be derived from the spa treatment suggested for gout (Section 14). When chronic albuminuria leads to anæmia, iron waters often act beneficially.

In the lesser forms of albuminuria, depending rather on a vice of the general nutrition than on any organic disease of the kidneys, spa treatment may be of use, according to individual indications, combined with an appropriate arrangement of diet and regimen.

In *congestion* of the kidneys from dilatation of the heart the treatment suggested under that heading ought to be applied (Section 35).

59. *Paroxysmal hæmoglobinuria* is not strictly an affection of the kidneys, nor is it suitable to balneo-therapeutic treatment, but rather to climatic influences; residence in warm non-malarious localities giving good results.

60. *Renal calculi*, of uric acid or oxalates, or of both, are occasionally passed during Karlsbad treatment, with or without the occurrence of renal colic. More frequently the tendency to fresh formation is checked by this treatment or by the alkaline waters of Vichy. Contrexéville and Wildungen, the principal types of the

Urinary calculi

earthy waters, are not rarely useful if they can be taken in large quantities—by an action which may be termed flushing out the kidneys. Often, however, the intended result is not obtained, either by these or by Karlsbad waters. As a preventive to calculous formation in the uric acid diathesis, the dietetic use of Luhatschowitz water (see p. 143), in doses of two or three tumblerfuls during the twenty-four hours, is often effective, and the risk of rendering the urine too alkaline, and thus favouring the deposition of phosphates, is avoided. The latter danger exists in the too long use of simple alkaline waters, though there are some people who can take for years a bottle of Vichy or Vals every day with advantage, and without their urine becoming alkaline. In plethoric persons a dose of 'bitter' or sulphated waters (Chapter XI.), twice or three times a week, is mostly beneficial, in addition to regular drinking of a tumberful of hot water night and morning, or of one of the slightly alkaline table-waters (Chapter XVI.).

Stone in the bladder requires surgical treatment, and this is also sometimes the case with calculus in the kidney, pelvis of the kidney, or ureter.

Pyelitis

61. *Chronic catarrh of the pelvis of the kidney* (chronic pyelitis), when not due to tubercle or an actual calculus, is, with proper diet and limitation of exercise, sometimes benefited by similar treatment to that recommended in Section 60; but the condition of the urine must be constantly watched. If the urine is apt to become alkaline, neither Karlsbad nor Vichy is permitted, though the waters of Contrexéville and Wildungen may often be used with advantage.

Vesical catarrh

62. The same may be said with regard to *chronic catarrh of the urinary bladder.*

Diseases and Disorders of the Sexual System.

(*a*) In *men* there are few affections of this class which are amenable to balneo-therapeutic treatment.

Sexual system in men

63. The diseases of the *testes* and the *prostate* belong much more to the sphere of ordinary medical and surgical treatment, and only associated derangements of health ought to come under spa treatment, and these ought to be attended to according to the suggestions already made under the proper headings.

A few words, however, may be devoted to affections of the *generative function*, for which advice is often demanded, and which are mostly connected with diminished sexual power, or more or less complete *impotency*. Frequently it happens that the want of power for which advice is asked is due to worn-out organs, either from general old age, or old age of the sexual organs. It is sometimes difficult to convince a man who is only fifty or sixty, or even less, that the power has ceased from premature old age, and that there are great differences in different men with regard to the period of life when these functions cease. It is useless to send such men to spas. Very often, however, the functions have not entirely ceased, and can be rendered again more active by strengthening waters and climate, and, above all things, by giving a long rest to the organs. In such cases of only partial or transitory impotency iron waters (Chapter XII.) and simple thermal waters (Chapter VI.) at high elevations have acquired a certain reputation, especially amongst the former, St. Moritz, amongst the latter, Gastein. Sea air and sea baths are likewise often useful. These remarks are also applicable to the cases of impotency from excess of sexual intercourse or from masturbation.

The temporary impotency caused by acute febrile diseases, such as severe enteric fever, is cured by 'time,' though the sexual power often does not return for many months. Tonic treatment by the ordinary remedies, by waters, and climate, are likely to be useful.

Bodily and mental breakdown from over-exertion and anxiety is likewise not rarely the cause of impotency, and the management in such cases is somewhat similar.

There are other cases in which virility is diminished or for a time entirely kept in abeyance by various morbid conditions; for instance, by dyspepsia with phosphaturia, by glycosuria, albuminuria, gout, dilatation of the heart, shock. In such cases the different causal conditions are to be treated either by ordinary means or by the spa treatment recommended under the respective headings, and if the cause be remedied the power will mostly return.

It is not necessary for us to discuss *sterility* in the male (aspermatism and azoöspermatism), for spa treatment can offer no cure.

(*b*) *Disorders of the Sexual System in Women.*

Amenorrhœa

64. *Disorders of menstruation.*—Of *amenorrhœa* only the varieties due to the general state of health can come under consideration for spa treatment; it is absolutely unnecessary to refer to cases due to imperfect local development. If the amenorrhœa is due to imperfect general development or to anæmia, one may, in addition to management of diet and regimen, recommend iron waters, long residence at the seaside with or without sea bathing, according to circumstances, or in mountain climates; but if dyspepsia and constipation are likewise present, the muriated waters

or the gaseous thermal muriated waters are likely to be beneficial. Time and patience are required, no over-treatment and no local interference should be permitted. If amenorrhœa is due to passive congestion of the womb, treatment similar to that for amenorrhœa from constipation may be recommended. Menstruation may be too scanty, or it may occur only at long intervals, and similar considerations to those just mentioned ought to guide the treatment. In cases combined with great torpor of the intestines and intestinal circulation, the waters of Franzensbad, combined with the moor baths of that place, are often very useful. In middle-aged women with amenorrhœa, combined with stoutness and sometimes with a rheumatic tendency, treatment at Franzensbad, Elster, or Marienbad is mostly of service.

In the management of deviations of the menstrual functions we have always to bear in mind that there are great natural varieties in different persons, varieties which are parts of individual habits and are within the limits of health, and these do not require interference any more by spa treatment than by ordinary treatment.

Dysmenorrhœa

65. In *dysmenorrhœa* likewise, some forms, such as mechanical or obstructive dysmenorrhœa, must be entirely excluded from consideration in this place. In the *congestive* form, with enlarged womb, often due to imperfect involution after confinement or abortion, the muriated waters (Chapter VII.) have obtained great reputation, and in some cases the muriated alkaline waters (Chapter IX.) have been found beneficial, but treatment in these cases ought not to be hurried; the usual course of three to four weeks is mostly insufficient, eight to ten weeks are often necessary, and a long stay at moderate elevations or at the seaside ought to follow. If the benefit

obtained by a course of four to six weeks is encouraging, but not sufficient, it will frequently be found useful to interrupt the spa treatment by a month spent at some not too distant climatic health resort of moderate elevation, in Switzerland or the Black Forest, and to resume it afterwards. The uterus absolutely requires a long rest.

In *ovarian dysmenorrhœa*, which is still more intractable, as an alternative to the means just mentioned, a long use of the simple thermal waters may be suggested. In very chronic cases of both forms (congestive and ovarian) the moor baths at Franzensbad often prove beneficial, as also the internal use of the Franzensbad springs.

Membranous dysmenorrhœa is perhaps the most obstinate form of dysmenorrhœa, but prolonged courses of treatment at the muriated alkaline and muriated spas, amongst which Ems and Baden-Baden have gained most adherents, or at the simple thermal spas, may be recommended with some chance of success.

Excessive menstruation. Menorrhagia

The *neuralgic form* is generally scarcely less obstinate. Here, too, the simple thermal spas may be recommended, and, if there is inactivity of bowels, Franzensbad.

66. *Excessive menstruation* may have different causes, and generally requires different kinds of home treatment, but when this fails it is in many instances advisable to have recourse to spa treatment, and not rarely long courses of chalybeate waters (Chapter XII.), and often the waters and baths of Franzensbad prove beneficial. A long rest at health resorts of moderate elevation ought always to follow the courses of waters.

Climacteric period

The menorrhagia depending on uterine fibroids will be shortly discussed under the latter head.

67. To the derangements of the general health, often connected with the *climacteric period*, we have already

alluded when speaking of disorders of the general health (Section 16). The sexual involution and the cessation of the menstrual functions are associated in many women with disorders of the function of the stomach and bowels and of the portal circulation, and of different spheres of the nervous system. The use of mineral waters under such circumstances requires greater caution than at other periods of life, but much benefit is often obtained from courses at Marienbad and Franzensbad when there is tendency to constipation and stoutness, and of the muriated waters (Chapter VII.), especially those of Homburg, Kissingen, and Baden-Baden in the case of lean individuals; Harrogate and Llandrindod are likewise often useful. A long stay at moderately elevated localities, or at the seaside, with rest from the social fatigues and worries of home life, is imperative in such cases. In delicate persons of neurotic tendency the simple thermal waters are preferable.

68. *Leucorrhœa* includes many varieties, and may be the result of different morbid conditions, general and local. The simplest varieties are *vulvar* and *vaginal leucorrhœa*, both more or less of the nature of catarrhs. If these affections do not yield to ordinary home treatment, muriated alkaline waters, such as those of Ems, can often be used with benefit; if constipation is associated with them, the muriated waters (Chapter VII.) are preferable, and in anæmic complications recourse may be had to iron waters (Chapter XII.). Well-arranged hydro-therapeutic treatment is likewise often useful. Tonic climates ought, if possible, to follow the course of waters.

Leucorrhœa

The *cervical*, and still more the *intra-uterine* and the *tubal* leucorrhœas, are much less adapted to balneo-therapeutic treatment, excepting in so far as the treatment can remove abdominal congestion (muriated waters, and amongst the alkaline sulphated, Franzensbad), or can do good by improving the condition of the blood (iron waters), or by allaying pain and hyperæsthesia (simple thermal and gaseous muriated thermal waters).

Diseases of the uterus and annexes

69. The majority of *diseases of the uterus* are not suitable to spa treatment. All acute conditions ought to be excluded, and not less so prolapse and other displacements, except if the latter are slight and due to a relaxed condition of the parts; in this latter case, in stout persons, the sulphated alkaline waters and moor baths of Franzensbad are sometimes attended with great success; in anæmic persons the internal use of iron waters with gaseous chalybeate baths or the muriated waters. In imperfect involution after confinements and abortions, in chronic endometritis, metritis, and perimetritis, in the remains of pelvic cellulitis from affections of the uterus, the cautious use of the muriated waters, such as Kreuznach, Woodhall Spa, Kissingen, Reichenhall, etc., and muriated alkaline waters, such as Ems, Royat, etc., with long rest of the affected parts, is frequently attended with benefit. Occasionally, after these classes of waters, the iron waters (Spa, Schwalbach, Pyrmont, etc.) are to be recommended for use as secondary courses.

Uterine fibroids

70. In fibroid tumours of the uterus, Kreuznach has enjoyed a great reputation. It is difficult to understand how muriated waters and baths can really exercise good effects on these tumours, but many unbiassed gynæcologists (amongst them the late Dr. Matthews Duncan) have assured us of great benefit, again and again derived by their patients from this treatment, especially by diminution of the menorrhagia. Such experience must be accepted, and we are inclined to

explain it by the improvement of the circulation in all the abdominal organs, including the uterus, and possibly some absorption of inflammatory products around the tumour. We have, however, never been informed of the complete disappearance of a fibroid tumour through spa treatment, previous to the entire cessation of menstruation.

71. *Tendency to miscarriage* is a subject for which women not rarely wish to have balneo-therapeutic advice. If the tendency is connected with anæmia and general debility, the waters recommended under these heads prove sometimes useful, but the treatment must be continued for a long time, and the womb, too, must under all circumstances have time to recover completely, before another pregnancy is risked.

Tendency to miscarriage

If the miscarriages are due to renal disease, courses of spa treatment can only in rare cases be of service.

If valvular disease with dilatation of the heart is the cause, the treatment recommended for the latter (Section 35) may be tried. At all events we have met with two cases where, after repeated courses of such treatment, and two years' rest for the uterus, the fœtus was retained to the natural term.

Whenever there is a well-founded suspicion of syphilis, this ought to be attended to, and we have then to decide either for home treatment or for specific treatment in combination with spa treatment.

Hysteria has been discussed under disorders of the nervous system.

Hysteria

72. *Sterility* in women is only in some conditions amenable to spa treatment. If the *ovaries* or the Fallopian tubes are imperfect or much diseased, or if the uterus or vagina is absent, it is useless to try cures by waters. Most affections of the uterus, too, such as

Sterility

defective development or displacements, cannot be improved by spa treatment. The limited amount which can be done by balneo-therapeutics in the chronic states of congestion or inflammation of the uterus and its lining and covering membranes, has been already alluded to in Section 69. Some waters, as the muriated alkaline (Chapter IX.), may assist in the cure of leucorrhœa and in correcting the acidity of the secretion by which the spermatozoa are destroyed; the cautious use of the ascending douche may, in addition, improve the circulation and nutrition of the uterus. It is well known that Ems has in such cases acquired a great reputation.

In spite of these considerations, by which the direct balneo-therapeutic action is so much restricted, we must acknowledge that we have numerous instances before us, in which well-arranged and long-continued spa and climatic treatment has been followed by pregnancy, and that a first pregnancy was in due time followed by further pregnancies without, and occasionally with the repeated help of balneo-therapeutic and climato-therapeutic treatment.

We have often been told that such results weaken or contradict our restrictions on the use of spa treatment in sterility, but a few further considerations may perhaps show that the good results mentioned, in some cases at least, allow other interpretations. We have seen such results follow the use of very different spas, such as Ems, Spa, Schwalbach, St. Moritz, Kissingen, Homburg, Franzensbad, Rippoldsau, Griessbach, Baden-Baden, Buxton, Plombières, St. Sauveur, Gastein, Wildbad, Ragatz. We have further seen them follow long seaside residence with and without sea bathing. Again, we have seen similar occurrences after long residence in the Alps, in the Black Forest, in the Pyrenees, in Egypt, and Algiers, without the use of any mineral waters, and also after long sea voyages.

We ought to add that, in almost all the cases alluded to, we had succeeded in inducing long separations of wife and husband, and it is to this circumstance that we are inclined to attribute much of the success. This interpretation, we may say, was shared by some of the physicians with whom we had consulted about the cases, and with whom we had later on discussed the results, especially the late Dr. Addison and Sir William Gull. The latter used to say that in the Middle Ages great ladies who had no children, and wished to have an heir, were often sent with their female attendants, but without their husbands, to distant shrines to pray for fertility, and that their prayers were not rarely granted, viz., that some time after their return they became pregnant.

In comparing such records with the results obtained by spas and climatic treatment, we are inclined to look at the matter somewhat in the following way. By the journey to the holy shrines and by the spa and climatic treatment, the health of the ladies had been improved, and this improvement was greatly aided by the powerful element of hope; by the long separation from the husbands and absence of excitement the sexual organs had enjoyed rest; the vigour on the part of the husband had been likewise increased, and thus the intercourse after the reunion became fruitful.

BIBLIOGRAPHY

Althaus, Julius, *The Spas of Europe.* London, 1862.
Baraduc, A., *Châtel-Guyon.* Paris, 1891.
Barnes, Robert, Contribution in *The Climates and Baths of Great Britain.* London, 1895.
Baumann, *Schlangenbad.* Wiesbaden, 1895.
Baumann, 'Die Wildbäder,' in Valentiner's *Handbuch.*
Beneke, F. W., *Ueber Nauheim's Soolthermen.* Marburg, 1859.
Beneke, F. W., *Weitere Mittheilungen über die Wirkungen der Soolthermen Nauheim's.* Marburg, 1861.
Beneke, F. W., 'Neue Erfahrungen über die Wirkungen der kohlensäurehaltigen Soolthermen Nauheim's.' *Berliner klin. Wochenschrift,* 1875, p. 109.
Beneke, F. W., *Zur Therapie des Gelenkrheumatismus und der ihm verbundenen Herzkrankheiten.* Berlin, 1872.
Blanc, Léon, *Rapport sur les Eaux Thermales d'Aix pendant l'Année* 1880. Paris, 1881.
Blanc, Léon, *Aix-les-Bains and Marlioz.* London, 1893.
Bottentuit, Eugène, 'Catarrhal Enteritis.' *Brit. Med. Journ.,* April 16, 1892.
Bottentuit, Eugène, *The Waters of Plombières.* London, 1888.
Bottey, F., *Traité Théorique et Pratique d'Hydrothérapie Médicale.* Paris, 1895.
Bowles, R. L., 'Nauheim and the Schott Treatment of Diseases of the Heart.' Proc. Harveian Society. *Lancet,* 1896, vol. i. p. 850.
Brabazon, A. B., 'Analysis of one hundred Cases of Rheumatoid Arthritis treated in the Royal Mineral Water Hospital, Bath.' *Brit. Med. Journ.,* 1896, vol. i. p. 723.
Brachet, *Aix-les-Bains.* London, 1884.
Brandt, G. H., *Royat,* 1883.
Braun, Julius, *On the Curative Effects of Baths and Waters.* English Edition, by Dr. Hermann Weber. London, 1875.
Braun, Julius, *Systematisches Lehrbuch der Balneotherapie.* Fifth Edition, by B. Fromm. Braunschweig, 1887.
Broadbent, J. F. H., and Broadbent, Sir William H., Bart., 'On the Treatment of Chronic Heart Disease by the Methods of Dr. Schott of Nauheim.' *Practitioner,* 1895.

Broadbent, Sir William H., Bart., 'Notes on Auscultatory Percussion and the Schott Treatment of Heart Disease.' *Brit. Med. Journ.*, 1896, vol. i. p. 769.

Brunton, T. Lauder, and Tunnicliffe, F. W., 'On the Effects of the Kneading of Muscles upon the Circulation, local and general.' *Journal of Physiology*, December 1894.

Brunton, T. Lauder, 'Atheroma and some of its Consequences, with their Treatment.' *Lancet*, October 12, 1895.

Cormack, C. E., *The Mineral Waters of Vichy.* London, 1887.

Dapper, Carl, 'Untersuchungen über die Wirkung des Kissinger Mineralwassers auf den Stoffwechsel des Menschen.' *Berl. klin. Woch.*, 1895. No. 31.

Débout-D'Estrées, *A Lecture on Contrexéville*, 1891.

Deetz, W., *Homburg vor der Höhe und seine Heilfactoren.* Second Edition, 1888.

Delastre, P., *Brides-les-Bains and Salins-Moutiers.* English Edition.

Delastre, P., *Les Albuminuriques aux Eaux de Brides-Salins* Paris, 1894.

Dickinson, W. H., *Treatise on Diabetes*, 1874.

Dickinson, W. H., Contribution in *The Climates and Baths of Great Britain.* London, 1895.

Diruf, Senior, Oscar, *Kissingen. Its Baths and Mineral Springs.* Würzburg, 1887.

Diruf, Senior, Oscar, 'Die Bitterwässer' in Valentiner's *Handbuch.*

Diruf, Senior, Oscar, and Niebergall, 'Die Kochsalzwässer' in Valentiner's *Handbuch.*

Doyon, A., *Uriage et ses Eaux Minérales.* Second Edition. Paris, 1884.

Dufresse de Chassaigne, J. E., *Mémoire sur le Traitement et la Guérison de l'Anévrysme rhumatismal du Cœur* (*i.e.* Rheumatic Valvular Disease) *sous l'Influence de l'Usage des eaux thermales de Bagnols.* Angoulême, 1859.

Durand-Fardel, Max., *Traité des Eaux Minérales.* Third Edition. Paris, 1883.

Durand-Fardel, Max., Eugène le Bret, J. Lefort and Jules François, *Dictionnaire Général des Eaux Minérales.* 2 vols. Paris, 1860.

Eardley-Wilmot, R., *On the Natural Mineral Waters of Leamington.* 1890.

Ebbesen and Hoerbye, *The Sulphureous Bath at Sandefjord in Norway.* (English Language). Christiania, 1862.

Eccles, A. Symons, 'On the Advantages of Oxidation.' *West London Medical Journal*, 1896, p. 4.

Égasse et Guyenot, *Eaux Minérales naturelles autorisées de France et de l'Algérie.* Second Edition. Paris, 1892.

Emond, E., *Le Mont-Dore et ses Eaux Minérales.* 1893.

Flechsig, Robert, *Bad Elster.* Third Edition. Leipzig, 1884.

Flechsig, Robert, *Bäder-Lexikon.* Second Edition. Leipzig, 1889.
Flechsig, Robert, *Handbuch der Balneotherapie.* Second Edition. Berlin, 1892.
Flechsig, Robert H., Balneological Notices in Schmidt's *Jahrbücher der gesammten Medicin.*
Forestier, Henri, *Le Traitement Thermal d'Aix-les-Bains.* Aix-les-Bains, 1895.
Fox, R. F., *Strathpeffer Spa, its Climate and Waters.* 1889.
Freeman, H. W., *The Thermal Baths of Bath.*
Frerichs, Th. von., *Ueber den Diabetes.* Berlin, 1884.
Friedlaender, R., *Beiträge zur Anwendung der physikalischen Heilmethoden.* Wiesbaden, 1896.
Fromm, B., Fifth Edition of Braun's *Lehrbuch der Balneotherapie.* 1887.
Garelli, G., *Acque Minerali d'Italia.* 1864.
Garrod, A. E., Contributions in *The Climates and Baths of Great Britain.* London, 1895.
Geissé, N., *The Springs of Ems.* Ems, 1892.
Genth, Carl, 'Ueber die Veränderung der Harnstoffausscheidung bei dem innerlichen Gebrauche des Schwalbacher Kohlensauren Eisenwassers,' in *Deutsch. Med. Wochenschrift.* 1887. No. 46.
Grossmann, 'Die alkalischen Quellen,' in Valentiner's *Handbuch.*
Gsell Fels, Th., *Die Bäder und klimatischen Kurorte der Schweiz.* Third Edition. Zürich, 1894.
Guentz, J. E., *Neue Erfahrungen über die Behandlung der Syphilis und Quecksilberkrankheit, mit besonderer Berücksichtigung der Schwefelwässer und Soolbäder.* Dresden, 1878.
De La Harpe, Eugène, *La Suisse Balnéaire et Climatérique.* Third Edition. Zürich, 1895.
De La Harpe, Eugène, *Formulaire des Eaux Minérales.* Second Edition. Paris, 1895.
De la Harpe, Eugène, *Louèche-les-Bains.* Paris, 1893.
Hayem, G., *Leçons de Thérapeutique. Les agents physiques et naturels.* Paris, 1894.
Head, Sir F. B., Bart., *Bubbles from the Brunnens of Nassau.* First Edition. London, 1834. This amusing work, published anonymously, although not strictly medical, is mentioned here on account of its great interest with regard to the spas of Schlangenbad, Schwalbach, etc.
Helfft, H., *Handbuch der Balneotherapie.* Ninth Edition. Edited by G. Thilenius. Berlin, 1882.
Hughes, Henry, 'Zur Wirksamkeit der Mineralbäder,' in *Deutsch. med. Woch.* 1893. Nos. 50–52.
Jacob, *Grundzüge der Balneotherapie.* 1870.
James, Constantin, and Aud'houi, Victor, *Guide Pratique aux Eaux Minérales.* Paris, 1893.
James, Prosser (jointly with C. R. C. Tichborne), *The Mineral Waters of Europe.* London, 1883.

Kennedy, Henry, *Observations on Fatty Heart.* Eighth Edition. Dublin, 1880.

Kerr, J. G. Douglas, *Popular Guide to the Use of the Bath Waters.* Ninth Edition.

Klein, Carl, *The Remedies of Franzensbad.* Franzensbad, 1889.

Lane, H., *General Hints on the Use of the Baths at Bath.* 1892.

Latham, P. W., *On Nervous or Sick Headache.* Cambridge, 1873.

Latham, P. W., Articles on 'Headache' and 'Megrim' in Quain's *Dictionary of Medicine.*

Leichtenstern, Otto, 'General Balneotherapeutics,' in Von Ziemssen's *Handbook of General Therapeutics.* English Translation by Dr. John Macpherson. London, 1885.

Leith, R. F. C., 'Action of Thermal Saline Baths and Resistance Exercises in the Treatment of Chronic Heart Disease.' *Lancet*, March 21 and 28, 1896.

Lersch, B. M., *Die phys. und therap. Fundamente der prakt. Balneologie.* 1868.

Liveing, Edward, *On Megrim, &c.* London, 1873.

Liveing, Robert, Articles on Skin Diseases in Quain's *Dictionary of Medicine.* Second Edition. 1894.

Ludwig, E., and Clar, C., *Ueber die Constantinquelle in Gleichenberg.* Wien, 1896.

Macé's *Guide aux Villes d'Eaux.* By various authors. First Edition. Paris, 1881.

Macpherson, John, *The Baths and Wells of Europe.* Third Edition. London, 1888.

Macpherson, John, *Our Baths and Wells: the Mineral Waters of the British Islands.* London, 1871.

Mapother, E. D., 'The Irish Sulphur Spas' in his *Papers on Dermatology.* London, 1889.

Meyer-Ahrens, C., *Die Heilquellen und Kurorte der Schweiz.* First Edition. Zürich, 1860.

Morris, Malcolm, Contributions in *The Climates and Baths of Great Britain.* London, 1895.

Müller, Franz C., The Balneological Notices in Schmidt's *Jahrbücher der gesammten Medicin*, from the year 1893 (after R. Flechsig's death).

Myrtle, A. S. and A., *Practical Observations on Harrogate Mineral Waters.* 1892.

Von Noorden, Carl, *Lehrbuch der Pathologie des Stoffwechsels.* Berlin, 1893.

Von Noorden, Carl, *Die Zuckerkrankheit und ihre Behandlung.* Berlin, 1895.

Von Noorden, Carl, 'On the Influence of the Salt Springs of Homburg, Kissingen, etc., on Metabolism in Man.' *The Practitioner*, London, March, 1896.

Oertel, M. J., 'Handbuch der allgemeinen Therapie der Kreislaufs-Störungen.' Fourth volume of Von Ziemssen's *Hand-*

buch der allgemeinen Therapie. Fourth Edition. Leipzig, 1891.
Oertel, M. J., *Ueber Terrain-Curorte zur Behandlung von Kreislauf-Störungen,* Leipzig, 1886.
Oliver, G., *The Harrogate Waters.* 1881.
Ord, W. M., Contributions in *The Climates and Baths of Great Britain.* London, 1895.
Pavy, F. W., *Researches on the Nature and Treatment of Diabetes.* Second Edition. London, 1869.
Pavy, F. W., *Physiology of the Carbo-hydrates.* 1894.
Penrose, F., Contributions in *The Climates and Baths of Great Britain.* London, 1895.
Petit, C. A., *Guide Médical aux Eaux de Royat.* Tenth Edition. Paris, 1896.
Pfeiffer, Emil, 'Thermal-Badecuren zu diagnostischen Zwecken.' *Berliner klin. Wochenschr.* 1896, p. 247.
Quain, Sir Richard, Bart., 'On Fatty Diseases of the Heart,' in *Med. Chir. Trans.* 1850.
Quain, Sir Richard, Bart., Lumleian Lecture on *Diseases of the Muscular Walls of the Heart.* 1872.
Quincke, H., *Balneolog. Tafeln.* 1872.
Rae, W. Fraser, *Austrian Health Resorts and the Bitter Waters of Hungary.* London, 1888.
Reimer, Hermann, *Handbuch der speciellen Klimatotherapie und Balneotherapie.* Berlin, 1889.
Von Renz, Wilh. Theodor, *Die Heilkräfte der sogenannten indifferenten Thermen, insbesondere bei Krankheiten des Nervensystems.* Tübingen, 1878.
Reumont, A., 'Die Schwefelquellen,' in Valentiner's *Handbuch.*
Reumont, A., *Die Thermen von Aachen und Burtscheid.* 1885.
Roberts, Frederick, Contributions in *The Climates and Baths of Great Britain.* London, 1895.
Robertson, W. H., 'The Medical Value of the Nitrogenous Tepid Water of Buxton.' *Lancet,* 1872 and 1874.
Robin, Albert, 'Des Albuminuries Phosphaturiques.' *Bulletin de l'Académie de Médecine.* Paris, 1893, p. 748.
Robin, Albert, 'Traitement Hydro-minéral des Albuminuries d'Origine Fonctionnelle ou Rénale.' *Annales d'Hydrologie.* Paris, 1896, p. 30.
Roehrig, *Die Physiologie der Haut.* Berlin, 1876.
Rotureau, A., Contributions on Mineral Waters in the *Dictionnaire Encyclopédique des Sciences Médicales.* Paris, 1864 to 1889.
Saundby, Robert, 'Remarks on the Nauheim (Schott) Treatment of Heart Disease.' *Brit. Med. Journ.,* November 2, 1895.
Savill, T. D., 'The Therapeutics of Saline Laxative Mineral Waters.' *Lancet,* November 23, 1895.
Schetelig, A., *Homburg.* 1893.

Scheuer, Victor, *Traité des Eaux de Spa.* Second Edition. Brussels, 1881.

Scheuer, Victor, *Essai sur l'action physiologique et thérapeutique de l'Hydrothérapie.* Paris, 1885.

Schivardi, Plinio, *Guida alle Acque Minerali ed ai Bagni d'Italia.* Fourth Edition. Milano, 1895.

Schott, August, 'Die Bedeutung d. Gymnastik f. d. Diagnose, Prognose u. Therapie d. Herzkrankheiten.' *Zeitschrift für Therapie.* 1885.

Schott, August und Theodor, 'Die Nauheimer Sprudel und Sprudelstrombäder.' *Berl. klin. Woch.*, 1884. No. 19.

Schott, Theodor, 'The Treatment of Chronic Diseases of the Heart by means of Baths and Gymnastics.' *Lancet*, 1890.

Schott, Theodor, *The Mineral Waters of Nauheim.* London, 1894.

Seegen, Joseph, *Handbuch der allgemeinen und speciellen Heilquellenlehre.* Second Edition. Vienna, 1862.

Seegen, Joseph, *Physiologisch-chemische Untersuchungen über den Einfluss des Glaubersalzes auf einige Factoren des Stoffwechsels.* Vienna, 1864.

Seegen, Joseph, *Die Zuckerbildung im Thierkörper.* Berlin, 1890.

Stoecker, A., *Bad Wildungen.* Fourth Edition. Edited by Dr. Marc. London, 1895.

Stoecker, A., 'Die erdigen Mineralquellen,' in Valentiner's *Handbuch.*

Sutro, Sigismund, *Lectures on the German Mineral Waters.* Second Edition. London, 1865.

Thilenius, G., Ninth Edition of Helfft's *Handbuch der Balneotherapie.* Berlin, 1882.

Thorne, W. Bezly. *The Schott Methods of the Treatment of Chronic Diseases of the Heart.* Second Edition. London, 1896.

Thorne, W. Bezly, 'Notes on certain Changes in the Cardio-vascular System which are induced by treatment according to the Schott Methods.' *Brit. Med. Journ.*, 1896, vol. i. p. 653.

Thorne, L. C. Thorne, 'Cases of Heart Disease treated by the "Schott" Method.' *Lancet*, January 4, 1896.

Tichborne, C. R. C., and James, Prosser, *The Mineral Waters of Europe.* London, 1883.

Tunnicliffe, F. W., Article with Dr. Brunton in the *Journal of Physiology*, December 1894.

Valentiner, Th., *Handbuch der allgemeinen und speciellen Balneotherapie.* Berlin, 1873.

Vintras, A., *Medical Guide to the Mineral Waters of France.* Second Edition. London, 1892.

Weber, Hermann, Article on 'Mineral Waters' in Quain's *Dictionary of Medicine.* Second Edition. London, 1894.

Weber, Hermann, English Edition of Braun's Work *On the Curative Effects of Baths and Waters.* London, 1875.

Weber, Hermann, The Marine Spas in 'The Treatment of Disease by Climate.' English translation by Dr. H. Port in Von

Ziemssen's *Handbook of General Therapeutics*, vol. iv., London, 1885.

Will, H., *Der Kurort Homburg vor der Höhe.* Homburg, 1881.

Will, H., *Dietetic and Therapeutic Hints to the Visitors of Bad Homburg.* Homburg, 1893.

Winternitz, W., 'Hydrotherapie' in Von Ziemssen's *Handbook of General Therapeutics.* English Translation by F. W. Elsner. London, 1886.

Wisard, A., *Traitement de l'Eczéma aux Eaux de St. Gervais.* 1895.

Wright, A. E., 'On the Treatment of Hæmorrhages and Urticarias, which are associated with Deficient Blood Coagulability.' *Lancet*, January 18, 1896.

Aachen als Kurort, by Doctors Alexander, Beissel, Brandis, Goldstein, Mayer, Rademaker, Schumacher, and Thissen. Edited by Dr. J. Beissel. Aachen, 1889.

Baden-Baden und seine Kurmittel, by Doctors Baumgärtner, von Corval, A. Frey, von Hoffmann, Schliep, Schneider. Baden-Baden, 1886.

The Climates and Baths of Great Britain. Being the Report of a Committee of the Royal Medical and Chirurgical Society of London. The contributions on the spas of Great Britain are by Doctors W. M. Ord, A. E. Garrod, F. Penrose, Robert Barnes, Frederick Roberts, and Mr. Malcolm Morris.

Deutsche Kurorte, edited by Dr. Oscar Lassar. Berlin, 1890.

Oeynhausen und seine Indicationen, edited, for the fifty years Jubilee of the spa, by Doctors Cohn, Huchzermeyer, Koch, Lehmann senior, Lehmann junior, Oetker, Reckmann, Rinteln, Rohden, Sauerwald, Voigt. Oeynhausen, 1895.

INDEX

The figures refer to the pages. The main references, when there are several, have been distinguished by thick figures.

Spottiswoode & Co. Printers, New-street Square, London.

www.ingramcontent.com/pod-product-compliance
Lightning Source LLC
Chambersburg PA
CBHW021355210326
41599CB00011B/881
9783337562168